Collective Light Emission

Many quantum emitters

Collective Light Emission

Many quantum emitters

Hsiang-Hua Jen
Institute of Atomic and Molecular Sciences, Academia Sinica, Taipei, Taiwan, Republic of China

IOP Publishing, Bristol, UK

ISBN 978-0-7503-2699-5 (ebook)
ISBN 978-0-7503-2697-1 (print)
ISBN 978-0-7503-2700-8 (myPrint)
ISBN 978-0-7503-2698-8 (mobi)

DOI 10.1088/978-0-7503-2699-5

Version: 20201001

IOP ebooks

British Library Cataloguing-in-Publication Data: A catalogue record for this book is available from the British Library.

Published by IOP Publishing, wholly owned by The Institute of Physics, London

IOP Publishing, Temple Circus, Temple Way, Bristol, BS1 6HG, UK

US Office: IOP Publishing, Inc., 190 North Independence Mall West, Suite 601, Philadelphia, PA 19106, USA

To my wife Yuwei for her unconditional love and wholehearted support,

and our two girls Michelle and Taryn for their enlightening laughter.

Contents

Preface

The subject of collective light emission presents one of the broadly investigated phenomena in quantum optics. After the laser, a coherent light source, came into existence, light–matter interacting physics advanced into a new era where quantum physics plays an essential role in many intriguing observations. This coherent light source shares many similarities with other later developed coherent forms of matter, for example the Bose–Einstein condensate.

It is the essence of collectivity in light emissions, arising from the coherent interactions between light and identical quantum emitters, that makes their properties distinct from those out of individual constituents. Many systems can exhibit this collective light emission. They can be as small as neutral atoms, ions, molecules, quantum dots, and as large as astronomical objects in the far distance. In addition, the temperature of these quantum systems can range from hundreds of Kelvin down to nanokelvin, respectively, in vapor cells or ultracold atoms under trapping potentials assisted by laser cooling technique. Finally, the spatial range of interaction mediated by light scattering within the system can be as long as infinite, for example in the guided modes of atom–nanofiber interfaces. This spatial range along with other cross-scale size and energy, make the light–matter interacting systems very complex. Therefore, they are one of the challenging physical platforms to tame.

Recently, much progress on light–matter interacting experiments and theories have emerged, owing to advanced experimental techniques which enable a better control over system parameters. This, in turn, inspired many theoretical proposals and fascinating ideas. In addition to the interests of fundamental studies, the efforts on generating efficient light–matter interfaces have also spurred the development of quantum information science. With flexible manipulations of system parameters at hand, quantum optics and condensed matter physics communities even ponder the possibility to quantum mechanically simulate exotic quantum states which are nonexistent in real materials. Understanding and keeping up with the subject of how light interacts with matter and the promise of using the light–matter interface as a quantum resource for quantum information processing gave birth to this book.

In this book, the basic mechanism of collective light emission and recent progress in theories and experiments are reviewed. Some significant predictions and observations are covered, along with unsettled issues of mismatch between observations and theoretical models. These will enlighten further studies in many versatile physical systems and pave the way towards full control and manipulation of collective light–matter interactions, en route to promising quantum supremacy from harnessing the capability of interacting light–matter interfaces.

Acknowledgements

I am grateful to my collaborators Y-C Chen, M-S Chang, and G-D Lin for many insightful discussions on most of the topics covered in this book. I am particularly indebted to my advisor and host S-K Yip during the funding period (2017–20) of the assistant research scholar by the Ministry of Science and Technology (MOST), Taiwan. The funding from MOST under the Grant Nos. MOST-106-2112-M-001-005-MY3 and MOST-109-2112-M-001-035-MY3, and the support of NCTS ECP1 (Experimental Collaboration Program) are well acknowledged.

Author biography

Hsiang-Hua Jen

Hsiang-Hua Jen received a bachelor degree in physics from National Taiwan University in 2000 and obtained his PhD degree in 2010 from Georgia Institute of Technology in the United States. He was trained in theoretical quantum optics and ultracold atoms and did his postdoc in National Tsing Hua University, Technische Universitt Kaiserslautern, and Institute of Physics of Academia Sinica (AS) in Taiwan. Now he actively investigates the collective effects of light–matter interactions, along with other interesting subjects of long-distance quantum communication and spinor Bose gases. In 2017, he was promoted to an assistant research scholar funded by MOST. Starting in March of 2020, he became an assistant research fellow at the Institute of Atomic and Molecular Sciences of AS. He has contributed to 30 peer-reviewed papers and was recently selected as 2019 IOP Outstanding Reviewer for *Journal of Physics B: Atomic, Molecular and Optical Physics*.

IOP Publishing

Collective Light Emission
Many quantum emitters
Hsiang-Hua Jen

Chapter 1

Introduction

Here I first introduce a brief history of quantum optics and its close relation with matter, which makes light–matter interaction one of the blossoming subjects explored in quantum physics. The essence of light–matter interactions lies in the collectivity emerging from the coherent interactions between the constituents within the medium, which is nontrivial in the phenomenon of collective light emissions and many other applications of quantum information processing and quantum computation.

1.1 How does light meet with matter?

The development of quantum optics can be traced back to Max Planck's theory of light quanta in 1900, when he investigated the spectral density of black-body radiation. He assumed that the energy of a harmonic oscillator in a cavity could be quantized and is associated with the frequency of the electromagnetic radiation. The idea is taken by Albert Einstein, who made one further step to explain the photoelectric effect, where the ejections of conduction electrons of a metal take place only by imparting sufficient quantized energy (frequency) of light. He received the prestigious Nobel prize in 1921 from this groundbreaking discovery, which lead to many insights on the quantum nature of light and matter and indispensable influences on the revolutions of quantum mechanics. The quantization of light energy is attributed as photons [1], which show distinct properties from classical light in light interferences and coherences [2, 3], or statistics on measurement.

Quantum optics thrived with the invention of a laser in the 1960s, along with the pioneering quantum theory of optical coherence by Roy J Glauber in 1963, who was later recognized in the 2005 Nobel physics prize. Since then, many subfields of quantum optics have emerged, which include nonlinear optics, atom optics, and quantum information processing, to name a few. The evolution of quantum optics is undoubtedly fascinating, in part due to the revolution of quantum physics in the early twentieth century, but mostly due to the intertwined relation in light–matter

doi:10.1088/978-0-7503-2699-5ch1

interactions [4]. In addition to advancing the quantum theory of light, for example, the phenomena of squeezed light, resonance fluorescence, laser theory, or quantum nondemolition measurements [5], the momentum transfer of light on matter and subsequent removal of energy accomplished a laser cooling technique and a subsequent achievement of Bose–Einstein condensate [6], which were awarded the 1997 and 2001 Nobel physics prizes, respectively. From the first realization of many-body ground states of a Bose–Einstein condensate in artificial setups, which is metastable and lives only for a couple of minutes in a laboratory, ultracold quantum gases offer an empowering paradigm to study and simulate many interesting and unique systems, for example, dipolar gases or polar molecules with long-range interactions, and superfluid to Mott insulator transitions in optical lattices. This shows the power of manipulations and the control of the laser technique in realizing the ground states or other exotic quantum phases in the setting of ultracold atoms.

Aside from the light-assisted realizations of condensed phases of matter, the individual quantum systems, say ions or Rydberg atoms, can as well be controlled and manipulated in their internal states. This further promises a versatile encoding and on-demand retrieval of quantum information that can be registered via light–matter couplings. The Nobel physics prize in 2012 awarded the experiments of ion in a trap and a photon in a cavity for pioneering experiments in controlling and measuring the atom by light in an ionic system or vice versa in the cavity QED system. From this recognition, we can see the essence of light–matter interactions, which enables interchanging roles between atoms and light via quantum measurement. Furthermore, the entanglement generated within light–matter interacting systems[1] can be transferred to far distances due to the nice features of light as information carriers. This leads to another active research realm of quantum communication[2].

When light meets with matter, especially in gaseous forms which we focus on in this book, its kinetic energy can be removed to reach a quantum regime. In this regime, the quantum fluctuations arise and low-temperature condensed matter physics emerges. On the other hand, the internal states of matter (hyperfine or Rydberg states) can also be accessed and manipulated by light excitations. The mechanism behind these manipulations of matter properties are closely related to light itself, which involves its central frequency, polarizations, linear and angular momenta. Without the concept of light quanta or a kind of corpuscular picture of light, many phenomena disclosed from isolated quantum systems cannot be explained. Of course a quantum particle should reveal a wave-particle duality, another essential feature in quantum physics, which complements the outcomes from either a wave- or particle-like fashion of light and atoms.

When light meets with matter, the 'lives' of them together are more intriguing than each with itself. One of the many fascinating phenomena in quantum optics

[1] The entanglement can be encoded in atomic (hyperfine ground states) or photonic (Fock space or polarization degrees), or even joint matter–light degrees of freedom.

[2] Quantum communication or long-distance quantum communication relies on transmitted photons in a telecommunication bandwidth, which has minimal transmission losses through optical fibers.

associated with light and matter together is collective light emissions from many quantum emitters, which we will cover in this book. This subject was extensively studied starting in the mid-twentieth century and now still raises and refreshes much interest in both theories and experiments. A marriage of them will always play important roles in renewing and advancing atomic, molecular, and optical physics, promising to give continuous insights into fundamental aspects of quantum physics, and stay close along the roadmap to many applications with new-generation quantum technology in the near future.

1.2 Collectivity that matters

In understanding our nature, we may take the approach of a reductionist's picture as a starting point. This approach intends to comprehend the world that we live in, including ourselves, less ambitiously, but gives us the first step towards a big picture of the whole. Though this big picture seems to be elusive in the near future, we still gain a lot by just trying. In physics, a single particle picture allows a comprehensive grasp of what happens in reality, and many uncomplicated cases can be explained simply by summing them up out of the results from individual constituents. This attributes to a noninteracting regime, where particles do not interact or only weakly, such that the properties of an ensemble of them (order of $\sim 10^{4-6}$ for cold atoms in an optical trap) or a usual matter of 10^{23} of them do not distinguish from the one of the single particle with only a linear scaling of number of particles. This seems to make life easier, but mother nature is more than complicated and cannot be comprehended only by this single particle picture.

It is the collectivity that makes all things different. Take animals as an example: many collective behaviors arise and show up in motions like fish schooling, bird flocking, or coordinated herds of wildebeest [7]. Life itself involves many complex mechanisms, and the motion of a group contrasting with a single entity can be at least due to the interaction between them. This can lead to unpredictability, fluctuations, multiple scales of dynamics, and emergent structures, which stops us from a complete comprehension of nature, but on the other hand enriches the forms and contents of it, which we would not perceive entirely for the moment. I will not specify what kind of interactions are mentioned here, since they depend on specific systems under study[3]. Therefore, a reductionist point of view in natural sciences merely provides small parts of the characteristics of a whole. This is similar to many scientific progresses where scientists with rigor work on the patches until a full grip of clear understanding. I will take Philip Anderson's own words 'more is different' [8], to sum up the complexity we encounter in explaining intricate phenomena or modeling systems we are interested, which states that a straightforward bottom-up approach is not able to reveal the essence of matter in mesoscopic or macroscopic scales. We should bear in mind that even an extra particle added, in an extreme sense, can make the whole quite different.

[3] For example, bosonic particles have short-range (hard-core) interactions, while a dipolar gas or electron has long-range dipole–dipole or Coulomb interactions, respectively.

When I talk about collectivity, I mean a mechanism that involves, in principle, all entities, which can be induced by collective interactions. From a Latin origin of collective, the 'collect' means 'gathered together'. Another usage on collective phenomena is 'cooperative', also adopted commonly in the physics community. It means 'worked together' for 'cooperat-' from Latin origin. I will not make a decisive or clear distinction between these two words here, for what we attribute to collective phenomena are actually outcomes from 'gathered workings'. For this reason, I will use these two usages interchangeably with no difference, and I hope this would not raise eyebrows here. To be honest, I expect no immediate settlement in the community on which is better or appropriate in describing the collective phenomena of light emissions for the moment.

In this book, I focus specifically on collective light emissions from many quantum emitters, which was originally coined as superradiance [9]. When treating these two-level emitters as spin-1/2 systems where the excited (ground) state is denoted as spin-up (down), a significant radiation intensity is proportional to N^2 with a number of particles N when the system is prepared as half excited. In huge contrast to the case of independent particles, the emitted light intensity is not linearly scaled by N as if they radiate individually. This is due to an emerging superradiant timescale which is shortened by N^{-1}, such that the light intensity magnifies as N^2. The result of superradiance by Robert H Dicke indeed surprises a lot from the first impression, since it suggests that a huge light intensity could be generated from exciting only a group of atoms. This seemingly surprising result is under the assumption of a light transition wavelength much larger than the ensemble size of these emitters, which leads to effectively structureless presumptions of atoms and thus is under a kind of ideal situation. As an estimate, I take rubidium atoms with D_2 transition wavelength of 780 nanometer ($5S_{1/2}$ to $5P_{3/2}$ levels) as the two-level quantum emitters. For just a number of 10^6 of them within say one tenth of the wavelength cubed, we obtain the number density as $\sim 2 \times 10^{27}$ cm^{-3} and a mass density as ~ 0.3 g cm^{-3}. This number approaches the average earth density of ~ 5 g cm^{-3}, and so the next question arises that why we do not observe superradiance all the time around us on Earth?

It turns out that we should be more careful in treating light–matter interactions, where light, absorbed and rescattered by atoms, travels at light speed within dilute gases, such that the macroscopic structure of atomic ensemble comes into play. Furthermore, the density of atoms cannot be made large indefinitely until the outer shell electrons 'touch' each other (radius size of rubidium atom ~ 3 Å), when quantum optical treatment fails and high energy physics has to step in. In figure 1.1, we demonstrate the rescattering events of light reabsorption and re-emission before radiating out of the whole medium. This leads to the resonant dipole–dipole interactions (RDDI) [10] in a form of r^{-3} and r^{-1} in the near and far field, respectively. This interaction form behaves exactly as two permanent dipoles, which depends on the dipole orientations and mutual distances, but in contrast, it is induced by external excitation fields.

In the left of figure 1.1, I particularly show one possible scattering path of light from deexcitation of one of the atoms in the medium. In real situations, the

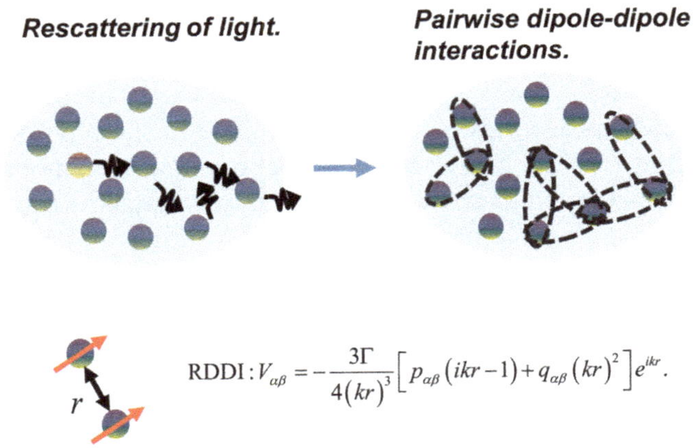

Figure 1.1. Collectivity arises from light rescattering within the atoms before emissions. The circle with dashes presents the pairwise resonant dipole–dipole interactions (RDDI), which connects with every other atoms via the functional form $V_{\alpha\beta}$. The interatomic distance is denoted as r, and $p_{\alpha\beta} \equiv \delta_{\alpha\beta} - (\hat{e}_\alpha^* \cdot \hat{r})(\hat{e}_\beta^* \cdot \hat{r})$, $q_{\alpha\beta} \equiv \delta_{\alpha\beta} - 3(\hat{e}_\alpha^* \cdot \hat{r})(\hat{e}_\beta^* \cdot \hat{r})$, with unit vectors of distance and dipole orientations $\hat{e}_\alpha$, and resonance wave vector k.

scattering process is more complicated, but can be simplified further by Markov and secular approximations typically used in quantum optics, which I will discuss in more detail in the next chapter. The end result is shown schematically in the right, where pairwise interactions show up in every other atoms of the whole ensemble, making all atoms strongly correlated via this long-range interaction $V_{\alpha\beta}$.

As a final remark in this section, we note of the requirement of identical quantum emitters in light–matter interacting systems. This relates to whether the systems can be entangled or not[4], and the quantum correlations can arise to differ from the classical ones. Distinguishable particles would show very different statistical properties from the identical ones, and so in the next section, I further introduce the concept of entanglement, in additional to the collectivity discussed here, which marks the boundary between classical and quantum worlds.

1.3 At the boundary from quantum to classical worlds

A very illustrative cartoon in an October issue of *Physics Today* in 1991 [11] delineates a live and dead cat (in dashed lines) in a quantum world on the other side of a gate, which separates from a classical realm where everything should be with certainty (drawing with solid lines). This is the so-called Schrödinger's cat, which intends to show the weirdness of quantum mechanics that seems to suggest an existing condition of an animal being both alive and dead.

It seems as if there is a paradox in interpreting quantum mechanical phenomena, but the paradox here actually arises when applying a correct theory to wrong objects. I do not mean that macroscopic objects cannot survive in a quantum world, but it really raises a lot more challenges to create such large quantum objects without

[4] In short, a bipartite system is called entangled when they cannot be expressed in product states. In other words, they can be distinguishable in the form of product states, and they are entangled otherwise.

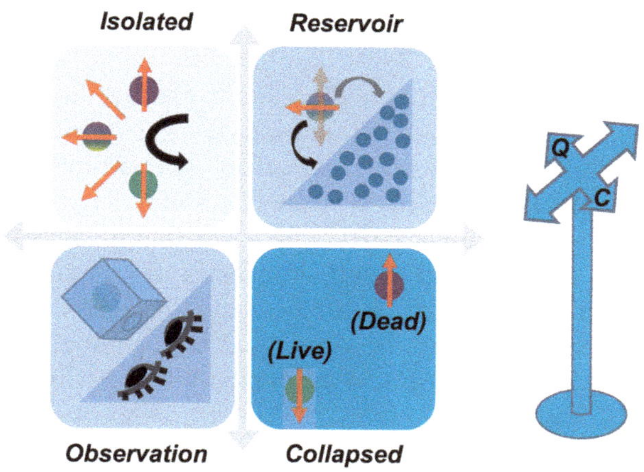

Figure 1.2. Four quadrants of a crossover from quantum to classical worlds with a sign stand. The isolated quadrant shows the clean interpretation of a quantum object (filled circles) with an intrinsic property of spin. Other than spin-up or spin-down in the quantized direction with solid colors, a tilted spin represents a superposition of the two. It collapses to a definite spin of reality in a classical realm as if downward (upward) spin is attributed to be live (dead) in essence of Schrödinger's cat. On collapsing to reality, a quantum object can go through reservoir interactions (energy exchange) or being observed by two bare eyes metaphorically. A sign stand presents the directions to quantum (Q) or classical (C) worlds.

being disturbed. Before going into detail of why these disturbances exist and where they come from, in figure 1.2 I focus on a true quantum object, if there is any, with an intrinsic spin denoted as straight arrows. This spin is quantized and only upward or downward spin are allowed (with solid colors attached to the objects), in contrast to a horizontal spin with mixed colors indicating an equal superposition of both spin configurations,

$$\frac{1}{\sqrt{2}}(|\uparrow\rangle + |\downarrow\rangle). \tag{1.1}$$

Therefore, the feature of superposition is one of the quantum mechanical property to describe a state of a quantum object. It seems as if there is no difference from tossing coins with a probability of one half for heads or tails, a statistical evidence for enough tosses. What differs is the coherence between these two spin configurations, and below I will further state the essential criteria to deviate from a classical reality.

In practical operations, a quantum system interacts with a reservoir that exchanges energy to and from the system. Sometimes the reservoir is referred to as environment, in a sense to show the influence from outside the quantum system. Effectively the reservoir decoheres the system of interest, and it represents a loss of information, leading to a decreasing entropy. Therefore, a quantum object always goes with a lifetime unless its information is stored in some kind of quantum memory with again a limited storage time. The state of a quantum object can also be monitored or observed, and then collapsed to reality of spin-up or spin-down. We show in figure 1.2 to demonstrate these two mechanism to transition from quantum to classical worlds. We note that a reservoir interaction is similar to a continuous observation, where a quantum particle can no longer be isolated whatsoever in a black box.

Collective light emission itself is a decoherent process, where initially excited atoms decay to the ground states with radiation. Its time duration is relevant to the lifetime of excited states, and the collective effect on this time scale emerges when many quantum emitters are involved. We here consider collectively excited quantum states that have no classical counterparts, where a distinction between them relies on whether these states are entangled or not. A measure of entanglement can attribute to von Neumann entropy, which is finite when the states are not able to be factorized, or otherwise they are separate and non-entangled. Other than the perspective of entanglement, I should also mention the vanishing macroscopic dipole moment in the Dicke's states of superradiance [9]. The superradiant phenomena can also be explained by product states, which are not entangled and possess a huge macroscopic dipole moment. This contrasts with superradiance originally proposed, and can be denoted as 'superfluorescence' [2] for a model of classical dipoles. In this book, I will follow the spirits of Dicke's states, which are highly entangled with a vanishing macroscopic dipole moment. These collective states will present the most profound theoretical predictions and are most relevant to many recent experimental observations.

To conclude this chapter, I put forward a promising future for quantum optical research. It will always find a place in many exciting fields of condensed matter, astrophysics, and material sciences, with advanced technical means developed continuously. This advancement along with theoretical developments can, for example some recent exciting achievements, potentially simulate quantum many-body states with long coherence time using optical microtraps [12], test gravitational effect on quantum correlations [13], and build a global quantum communication via satellite-based channels [14–16]. For a subfield like collective light emission, I expect that it will remain an active research direction for its universality in many convenient and versatile platforms constructed worldwide. The characteristics of collectivity and quantum correlations in light emissions will give insight to fundamental studies of light–matter interfaces and offer many opportunities in quantum information sciences that will have a huge impact on human history.

References

[1] Loudon R 2000 *The Quantum Theory of Light* 3rd edn (Oxford: Oxford University Press)

[2] Mandel L and Wolf E 1995 *Optical Coherence and Quantum Optics* (Cambridge: Cambridge University Press)

[3] Scully M O and Zubairy M S 1997 *Quantum Optics* (Cambridge: Cambridge University Press)

[4] Cohen-Tannoudji C, Dupont-Roc J and Grynberg G 1992 *Atom-Photon Interactions: Basic Processes and Applications* (New York: Wiley)

[5] Walls D F and Milburn G J 1995 *Quantum Optics* (Berli: Springer)

[6] Pethick C J and Smith H 2002 *Bose-Einstein Condensation in Dilute Gases* (Cambridge: Cambridge University Press)

[7] Katherine W 2019 Life is physics *Am. Phys. Soc. Phys.* **12** 2

[8] Anderson P W 1972 *More is different Science* **177** 393–6
[9] Dicke R H 1954 Coherence in spontaneous radiation processes *Phys Rev.* **93** 99–110
[10] Lehmberg R H 1970 Radiation from an N-atom system. I. General formalism *Phys. Rev.* A **2** 883–8
[11] Zurek W H 1991 Decoherence and the transition from quantum to classical *Phys. Today* **10** 36–44
[12] Norcia M A, Young A W, Eckner W J, Oelker E, Ye J and Kaufman A M 2019 Seconds-scale coherence on an optical clock transition in a tweezer array *Science* **366** 93–7
[13] Xu P *et al* 2019 Satellite testing of a gravitationally induced quantum decoherence model *Science* **366** 132–5
[14] Scheidl T, Wille E and Ursin R Quantum optics experiments using the International Space Station: a proposal *New J. Phys.* **15** 043008
[15] Yin J *et al* 2017 Satellite-based entanglement distribution over 1200 kilometers *Science* **356** 1140–4
[16] Liao S K *et al* 2018 Satellite-relayed intercontinental quantum network *Phys. Rev. Lett.* **120** 030501

Chapter 2

Light-induced dipole–dipole interactions

The collective and resonant dipole–dipole interactions arise in the dissipation process induced by common light excitations. They are pairwise for every other atoms in the medium, reflecting a highly correlated scattering process in excitations and de-excitations of quantum emitters. Here I show the foundations for resonant dipole–dipole interactions in Heisenberg equations of motion and obtain very distinct long-range forms of interactions in free space, and two- and one-dimensional reservoirs.

2.1 Reservoir interaction volume

Before showing the main results of resonant dipole–dipole interactions induced in the dissipation process, I first present the concept of reservoir interaction volume in figure 2.1. The typical quantum theory in treating spontaneous emissions are based on atoms interacting with bosonic fields from reservoirs [1]. These reservoir volumes in various dimensions are actually quantization volumes of bosonic fields allowed in systems of interests. Figure 2.1 schematically demonstrates coordinates used to characterize the degrees of freedom of bosonic fields, and the examples of atoms in various dimensions are shown.

In free space, it can be easily understood that a trapped atomic ensemble presents one of the platforms that mediate three-dimensional (3D) reservoir bosonic modes. Two- (2D) and one-dimensional (1D) reservoirs are more tricky to design, since they do not naturally exist. In the below, I will address how these lower dimensional reservoirs can be devised and realized, and the distinctive features of dissipations emerge for different allowable dimensions of reservoirs.

2.2 Resonant dipole–dipole interactions in free space

Collective spontaneous emissions show very different timescales due to the resonant dipole–dipole interactions induced in the dissipation process. Here I first introduce the origin of induced dipole–dipole interactions of systems in free space reservoirs.

doi:10.1088/978-0-7503-2699-5ch2

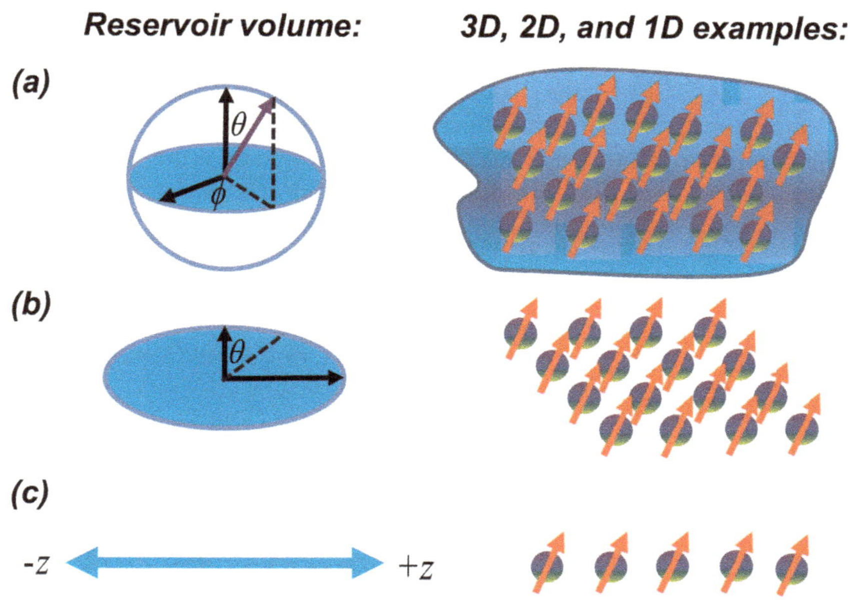

Figure 2.1. Reservoir volumes and examples of three- (3D), two- (2D), and one-dimensional (1D) systems. The reservoir interaction volume denotes the quantization volume of bosonic fields that interact with the systems of interests. Respectively, (a) spherical coordinates with θ and ϕ, (b) polar coordinates with θ, and (c) the one-dimensional reservoir allowed in $\pm z$ directions, demonstrate the degrees of freedom in various quantization volumes. Corresponding examples of atomic ensembles are shown schematically, associated with uniform spins induced by light polarizations.

This pairwise dipole–dipole interaction between every other quantum emitter can be obtained from a generic Hamiltonian [2] for atom-reservoir interactions,

$$H = \sum_{\mu=1}^{N} \hbar\omega_e \hat{\sigma}_\mu^\dagger \hat{\sigma}_\mu - \sum_{\mu=1}^{N} \sum_{q,\lambda} g_q \left(\vec{\epsilon}_{q,\lambda} e^{i\mathbf{k}_q\cdot\mathbf{r}_\mu - i\omega_q t} \hat{a}_q + \vec{\epsilon}^{\,*}_{q,\lambda} e^{-i\mathbf{k}_q\cdot\mathbf{r}_\mu + i\omega_q t} \hat{a}_q^\dagger \right) \cdot \hat{d} \left(\hat{\sigma}_\mu + \hat{\sigma}_\mu^\dagger \right). \quad (2.1)$$

The Hamiltonian H is written down under the dipole approximation [1], where the size of a bounded electron is much smaller than the transition wavelength, such that the light–matter interaction can be simply described by the dipole energy of $-\mathbf{d} \cdot \mathbf{E}$, with an electric field $\mathbf{E}$ and a dipole $\mathbf{d}$. Here I consider N two-level quantum emitters of alkali atoms with $|g\rangle$ and $|e\rangle$ for the ground and excited states respectively. The transition energy is $\hbar\omega_e$, and the atomic dipole operator is $\hat{\sigma}_\mu^\dagger \equiv |e\rangle_\mu\langle g|$ with $\hat{\sigma}_\mu = (\hat{\sigma}_\mu^\dagger)^\dagger$. The quantized fields $\hat{a}_q$ should satisfy the bosonic commutation relations $[\hat{a}_q, \hat{a}_{q'}^\dagger] = \delta_{q,q'}$, which represent a reservoir of quantized bosonic light fields exchanging energy with atoms. The coupling constant is denoted as $g_q \equiv d/\hbar\sqrt{\hbar\omega_q/(2\epsilon_0 V)}$, which involves a dipole moment d with its unit direction $\hat{d}$, two allowable polarizations of the fields $\vec{\epsilon}_{q,\lambda=1,2}$ of modes q, and a quantization volume V. The spatial distributions of the quantum emitters are described by $\mathbf{r}_\mu$.

We note that the Hamiltonian in equation (2.1) is written without applying the rotating-wave approximations, and this is necessary for an appropriate treatment on the frequency shift (dispersion) of the dipole–dipole interactions. Without the rotating-wave approximations that throw away the non-rotating wave terms, the dispersion and absorption of the interactions should satisfy the Kramers–Kronig relation [3]. This seems to be in contrast to many other quantum optical settings with the rotating-wave approximations, but we should be aware that the approximation is valid only when the so-called virtual process in an energy scale of $(\omega_e + \omega_q)$ (atomic excitation along with a photon emission) does not contribute to physical observables. It turns out that in obtaining a correct frequency shift from the resonant dipole–dipole interactions, the non-rotating wave terms should be kept intact [4, 5] and the Kramers–Kronig relation sustains.

2.2.1 Heisenberg equations and Lindblad forms

To obtain self-consistent coupled equations between the atoms and the reservoir bosonic fields, we proceed to use the Heisenberg equation $d\hat{Q}/dt = i[H, \hat{Q}]$ (let $\hbar = 1$) for an arbitrary atomic operator $\hat{Q}$. We obtain

$$\begin{aligned}\frac{d\hat{Q}}{dt} = i\omega_e \sum_\mu \left[\hat{\sigma}_\mu^\dagger \hat{\sigma}_\mu, \hat{Q}\right] - i \sum_\mu \sum_{q,\lambda} g_q \Big\{ e^{i\mathbf{k}_q\cdot\mathbf{r}_\mu}\left[\hat{\sigma}_\mu + \hat{\sigma}_\mu^\dagger, \hat{Q}\right](\vec{\epsilon}_{q,\lambda}\cdot\hat{d})\hat{a}_q(t) \\ - e^{-i\mathbf{k}_q\cdot\mathbf{r}_\mu}\left(\vec{\epsilon}^{\,*}_{q,\lambda}\cdot\hat{d}\right)\hat{a}_q^\dagger(t)\left[\hat{Q}, \hat{\sigma}_\mu + \hat{\sigma}_\mu^\dagger\right]\Big\},\end{aligned} \tag{2.2}$$

where $\hat{a}_q$ can be substituted by solving $d\hat{a}_q/dt = i[H, \hat{a}_q]$,

$$\hat{a}_q(t) = \hat{a}_q(0)e^{-i\omega_q t} + i\sum_\mu g_q(\vec{\epsilon}_{q,\lambda}\cdot\hat{d})e^{-i\mathbf{k}_q\cdot\mathbf{r}_\mu}\int_0^t dt'\left[\hat{\sigma}_\mu(t') + \hat{\sigma}_\mu^\dagger(t')\right]e^{-i\omega_q(t-t')}. \tag{2.3}$$

We further apply the Born–Markov approximation, where $\omega_e t \gg 1$ and $t \gg (r_{\mu\nu})_{\max}/c$ ($r_{\mu\nu} \equiv |\mathbf{r}_\mu - \mathbf{r}_\nu|$ and max for maximal mutual distance) assuming no memory effect of interactions, we obtain the time evolutions of $Q \equiv \langle\hat{Q}\rangle_0$ in Lindblad forms by considering the vacuum bosonic fields $\langle\rangle_0$,

$$\dot{Q}(t) = \sum_{\mu\neq\nu} i\Omega_{\mu,\nu}\left[\sigma_\mu^\dagger\sigma_\nu, Q\right] + \mathcal{L}(Q), \tag{2.4}$$

$$\mathcal{L}(Q) = \sum_{\mu,\nu}\gamma_{\mu,\nu}\left[\sigma_\mu^\dagger Q\sigma_\nu - \frac{1}{2}\left(\sigma_\mu^\dagger\sigma_\nu Q + Q\sigma_\mu^\dagger\sigma_\nu\right)\right]. \tag{2.5}$$

The $\Omega_{\mu,\nu}$ and $\gamma_{\mu,\nu}$ represent the collective frequency shifts and decay rates, respectively, of the pairwise couplings, $J_{\mu,\nu} \equiv (\gamma_{\mu,\nu} + i2\Omega_{\mu,\nu})/2$, which we define as

$$
\begin{aligned}
J_{\mu,\nu} &= \sum_{q,\lambda} |g_q|^2 |\vec{\epsilon}_{q,\lambda} \cdot \hat{d}|^2 \int_0^\infty dt' e^{i\mathbf{k}_q\cdot(\mathbf{r}_\mu - \mathbf{r}_\nu)}[e^{i(\omega_e-\omega_q)t'} + e^{-i(\omega_e+\omega_q)t'}], \\
&= \sum_{q,\lambda} |g_q|^2 |\vec{\epsilon}_{q,\lambda} \cdot \hat{d}|^2 e^{i\mathbf{k}_q\cdot(\mathbf{r}_\mu - \mathbf{r}_\nu)}\big[\pi\delta(\omega_q - \omega_e) + \pi\delta(\omega_q + \omega_e) \\
&+ i\mathcal{P}(\omega_e - \omega_q)^{-1} - i\mathcal{P}(\omega_q + \omega_e)^{-1}\big],
\end{aligned} \tag{2.6}
$$

where $\mathcal{P}$ is the principal value of the integral. From the above, we can identify the non-rotating wave terms $\propto (\omega_q + \omega_e)$ which are neglected in most of other quantum optical models, but they are kept here for consistency.

For a reservoir in free space, we consider continuous limits of modes $\sum_q \to \int_{-\infty}^{\infty} \frac{V}{(2\pi)^3} d^3q$. Considering a spherical coordinate, we obtain $J_{\mu,\nu}$ in free space [2],

$$
\begin{aligned}
\gamma_{\mu,\nu}(\xi) &\equiv \oint d\Omega_q [1 - (\hat{\mathbf{q}} \cdot \hat{\mathbf{p}})^2] \int_0^\infty dq q^2 g_q^2 \frac{V}{(2\pi)^3}\big[\pi\delta(\omega_q - \omega_e) + \pi\delta(\omega_q + \omega_e)\big], \\
&= \frac{3\Gamma}{2}\left\{\big[1 - (\hat{\mathbf{p}} \cdot \hat{r}_{\mu\nu})^2\big]\frac{\sin\xi}{\xi} + \big[1 - 3(\hat{\mathbf{p}} \cdot \hat{r}_{\mu\nu})^2\big]\left(\frac{\cos\xi}{\xi^2} - \frac{\sin\xi}{\xi^3}\right)\right\},
\end{aligned} \tag{2.7}
$$

$$
\begin{aligned}
\Omega_{\mu,\nu}(\xi) &\equiv - \oint d\Omega_q [1 - (\hat{\mathbf{q}} \cdot \hat{\mathbf{p}})^2] \int_0^\infty dq q^2 g_q^2 \frac{V}{(2\pi)^3} \\
&\quad \big[i\mathcal{P}(\omega_q - \omega_e)^{-1} + i\mathcal{P}(\omega_q + \omega_e)^{-1}\big], \\
&= \frac{3\Gamma}{4}\left\{-\big[1 - (\hat{\mathbf{p}} \cdot \hat{r}_{\mu\nu})^2\big]\frac{\cos\xi}{\xi} + \big[1 - 3(\hat{\mathbf{p}} \cdot \hat{r}_{\mu\nu})^2\big]\left(\frac{\sin\xi}{\xi^2} + \frac{\cos\xi}{\xi^3}\right)\right\},
\end{aligned} \tag{2.8}
$$

where $d\Omega_q$ denotes a solid angle integration, $\hat{\mathbf{p}}$ parallels the excitation field polarization, the natural decay constant $\Gamma = d^2\omega_e^3/(3\pi\hbar\epsilon_0 c^3)$, and a dimensionless length scale $\xi \equiv k_L|\mathbf{r}_\mu - \mathbf{r}_\nu|$ with $k_L = \omega_e/c$. The $\Gamma \sim 2\pi \times 6$ MHz for rubidium atoms, for example, leading to a timescale of 26 ns, and ξ denotes a measure of interatomic distance that characterizes the strength of interactions. For $\xi \gg 1$, the dominating term in $J_{\mu,\nu}$ becomes $\propto e^{-i\xi}/\xi$.

The above results can be obtained by assuming a dipole orientation in the x–z plane with a polar angle θ' from the z-axis and a wave vector mode q with a general orientation (θ, ϕ) as polar and azimuthal angles in a solid angle integration of Ω_q. The results of equations (2.7) and (2.8) can be expressed further in compact forms as shown in figure 1.1, where the induced dipole–dipole interactions act as if two permanent dipoles interact with each other. The oscillatory features of the interaction forms can be seen in figure 2.2, which can last at long distances between any two atoms. The feature of long-range interactions can be further seen in logarithmic scales in the inset of figure 2.2, where even one thousandth interaction strength still sustains at $\xi \gtrsim 20$. Since typical atomic ensembles involve more than hundreds of thousand atoms, these long-range and pairwise dipole–dipole interactions can be collectively enhanced in macroscopic scales of dense atoms.

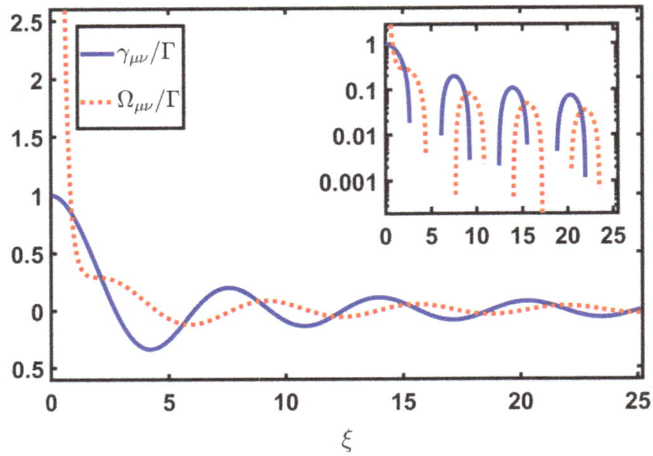

Figure 2.2. Oscillatory resonant dipole–dipole interaction forms. Here an example of $\mathbf{p} \perp r_{\mu\nu}$ for $\gamma_{\mu\nu}$ and $\Omega_{\mu\nu}$ is demonstrated, where oscillatory behaviors last up to $\xi = 25$. The inset further shows logarithmic scales of positive $\gamma_{\mu\nu}$ and $\Omega_{\mu\nu}$ (negative parts are ignored).

We note of the divergence in frequency shifts $\Omega_{\mu\nu}$ as $\xi \to 0$. This shows unphysical predictions and a failure of quantum optical treatments in such a short distance between atoms. The size of radius of rubidium atoms is about 0.25 nanometers (nm), which delineates a limit of allowable atomic separation when their electronic orbitals touch each other, that is $\xi \sim 4 \times 10^{-3}$ for $\lambda = 780$ nm. Comparing a density of Bose–Einstein condensate $\rho \sim 10^{14}$ cm^{-3}, an average atomic separation reads $\sqrt[3]{\rho^{-1}} \sim 0.28\lambda$, which leads to $\xi \sim 1.7$. Therefore, a ξ of an order of one is about what we can achieve in a dense and macroscopic atomic gas, and $\xi \sim 0.1$ presents a strong interaction limit with significant frequency shifts and decay rates. For $\xi \sim 0.01$, close to the regime where quantum optics fails, we should be careful on interpreting the results in figure 2.2 especially for $\Omega_{\mu\nu}$. Though it seems a strong interaction limit of $\xi \sim 0.1$ is hard to reach in gaseous atoms, we note that other possible platforms of quantum emitters, for example of superconducting qubits with microwave transitions or quantum dots, can in principle get into a strong interaction regime more easily and show a more enhanced collective effect of emissions. As a final remark, we present the resonant dipole–dipole interactions from the quantum emitters with a two-level configuration here, while an extension to the case of the atoms with multiple atomic levels (strontium atoms for example) is straightforward [6].

2.3 Two-dimensional dipole–dipole interactions

Next we extend the results in the previous section and allow a reservoir interaction only in a flat plane. This 2D dipole–dipole interaction is of particular interest in recently more accessible two-dimensional devices of coupled ring resonators [7]. Therefore, it opens up new revenues to explore collective emissions of quantum emitters in a 2D reservoir. This specially designed 2D reservoir can be realized in, for example, 2D optical lattices of atoms inside a planar cavity with perfect mirrors or

2D arrays of superconducting qubits. We note that a similar discussion of 2D electric dyadic Green's function has been carried out with a 2D source current in a dielectric [8], and scattering from 2D point scatterers [9], and tailoring of superradiance in various dimensional electromagnetic reservoirs [10] has been investigated. Below we present the essential results of 2D dipole–dipole interaction forms [11] and make a direct comparison to the case in free space in the previous section.

From equation (2.6), we consider a 2D reservoir with a quantization area A for N two-level atoms with arbitrary 2D geometries. Under polar coordinates, we obtain the 2D version of $J_{\mu,\nu}$,

$$\begin{aligned} J_{\mu,\nu} = & \int_0^\infty \frac{qg_q^2 A}{(2\pi)^2} dq \int_0^{2\pi} d\theta [1 - (\hat{\mathbf{q}} \cdot \hat{\mathbf{p}})^2] e^{i\mathbf{k}_q \cdot \mathbf{r}_{\mu,\nu}} \\ & \left[\pi\delta(\omega_q - \omega_e) + \pi\delta(\omega_q + \omega_e) + i\mathcal{P}(\omega_e - \omega_q)^{-1} - i\mathcal{P}(\omega_q + \omega_e)^{-1}\right], \end{aligned} \tag{2.9}$$

where $\mathbf{r}_{\mu,\nu} = \mathbf{r}_\mu - \mathbf{r}_\nu$, and $\hat{\mathbf{p}}$ denotes the excitation polarization. We first calculate the real part of $J_{\mu,\nu}$, that is the dissipative part $\gamma_{\mu,\nu}$, and denote a general $\hat{\mathbf{q}}$ with a polar angle θ to the $\hat{y}$. Without loss of generality, we choose $\mathbf{r}_{\mu,\nu}$ along $\hat{y}$ and $\hat{\mathbf{p}}$ with a polar angle θ' to $\hat{y}$. We then obtain

$$\begin{aligned} \frac{\gamma_{\mu,\nu}}{2} &= \int_0^\infty \frac{qg_q^2 A}{(2\pi)^2} dq \int_0^{2\pi} d\theta [1 - (\cos\theta\cos\theta' + \sin\theta\sin\theta')^2] e^{i\xi\cos\theta} \pi\delta(\omega_q - \omega_e), \\ &= \Gamma_{2D} \frac{1}{\pi} \int_0^{2\pi} d\theta [1 - (\cos\theta\cos\theta' + \sin\theta\sin\theta')^2] e^{i\xi\cos\theta}, \end{aligned} \tag{2.10}$$

where the intrinsic decay constant for a 2D reservoir is

$$\Gamma_{2D} \equiv \frac{1}{4} k_L |\partial_\omega q(\omega)|_{\omega=\omega_e} g_{k_L}^2 A, \tag{2.11}$$

with a 2D coupling strength gk_L and the inverse group velocity $\partial_\omega q(\omega)$. We proceed to obtain the explicit expression by integrating out the polar angles,

$$\begin{aligned} \frac{\gamma_{\mu,\nu}}{2} &= 2\Gamma_{2D}\left[J_0(\xi) - \frac{J_1(\xi)}{\xi} + (\hat{\mathbf{p}} \cdot \hat{\mathbf{r}}_{\mu,\nu})^2 J_2(\xi)\right] \\ &\equiv \Gamma_{2D} f(\xi), \end{aligned} \tag{2.12}$$

where a unit direction $\hat{\mathbf{r}}_{\mu,\nu} = (\mathbf{r}_\mu - \mathbf{r}_\nu)/|\mathbf{r}_\mu - \mathbf{r}_\nu|$. The above results can be derived from the following integrals,

$$\int_0^{2\pi} e^{ia\cos\theta} d\theta = 2\pi J_0(|a|), \tag{2.13}$$

$$\int_0^{2\pi} \cos^2\theta e^{ia\cos\theta} d\theta = 2\pi\left(\frac{J_1(a)}{a} - J_2(a)\right), \tag{2.14}$$

$$\int_0^{2\pi} \sin^2\theta \, e^{ia\cos\theta} \, d\theta = 2\pi \frac{J_1(|a|)}{|a|}, \tag{2.15}$$

$$\int_0^{2\pi} \sin\theta \cos\theta \, e^{ia\cos\theta} \, d\theta = 0, \tag{2.16}$$

where $J_n(a)$ are the Bessel functions of the first kind.

The $\gamma_{\mu,\nu}$ and $\Omega_{\mu,\nu}$, real and imaginary parts of $J_{\mu,\nu}$ respectively, should satisfy the Kramers–Kronig relation, which is required for causality in a physical response function [3]. Finally we obtain $J_{\mu,\nu}$,

$$\begin{aligned} J_{\mu,\nu} &= \Gamma_{2D} f(\xi) - \frac{i\mathcal{P}}{2\pi} \int_0^\infty d\omega \Big(k_L |\partial_\omega q(\omega)| \bar{g}_q^2 A/2 \Big) f(\omega |\mathbf{r}_\mu - \mathbf{r}_\nu|/c) \\ &\quad \left(\frac{1}{\omega - \omega_e} + \frac{1}{\omega + \omega_e} \right), \\ &= \Gamma_{2D}[f(\xi) + ig(\xi)], \end{aligned} \tag{2.17}$$

where

$$f(\xi) \equiv 2\left[J_0(\xi) - \frac{J_1(\xi)}{\xi} + (\hat{\mathbf{p}} \cdot \hat{\mathbf{r}}_{\mu,\nu})^2 J_2(\xi) \right], \tag{2.18}$$

$$g(\xi) \equiv 2Y_0(\xi) - 2\frac{Y_1(\xi)}{\xi} + 2(\hat{\mathbf{p}} \cdot \hat{\mathbf{r}}_{\mu,\nu})^2 Y_2(\xi) - \frac{4}{\pi\xi^2}\left[1 - 2(\hat{\mathbf{p}} \cdot \hat{\mathbf{r}}_{\mu,\nu})^2\right], \tag{2.19}$$

and $Y_n(\xi)$ are the Bessel functions of the second kind. The above functional form of $g(\xi)$ can be derived by using the following integrals,

$$\mathcal{P}\int_0^\infty da \frac{J_0(a)}{a \mp b} = -\frac{\pi}{2}[Y_0(b) \pm H_0(b)], \tag{2.20}$$

$$\mathcal{P}\int_0^\infty da \frac{J_1(a)}{a(a \mp b)} = -\frac{2 + \pi b[Y_1(b) \pm H_1(b)]}{2b^2}, \tag{2.21}$$

$$\mathcal{P}\int_0^\infty da \left(\frac{J_2(a)}{a - b} + \frac{J_2(a)}{a + b} \right) = -\frac{4}{b^2} - \pi Y_2(b), \tag{2.22}$$

where $H_n(b)$ is the Struve function.

Similar to the case in free space, the coherent and dissipative coupling forms of $g(\xi)$ and $f(\xi)$, respectively, denote the collective frequency shifts and decay rates between any two atoms. In figure 2.3, 2D dipole–dipole interactions of equation (2.17) are shown for two polarizations of light with $\hat{\mathbf{p}} \parallel \hat{\mathbf{r}}_{\mu,\nu}$ and $\hat{\mathbf{p}} \perp \hat{\mathbf{r}}_{\mu,\nu}$, respectively.

For a short range, both $\gamma_{\mu,\nu} \propto f(\xi)$ approach unity, which are the same as in Dicke's superradiant regime. For $\Omega_{\mu,\nu} \propto g(\xi)$ in a short range, the asymptotic function is $\sim\ln(\xi/2)$, which is non-analytic as $\xi \to 0$, but has a much less divergent

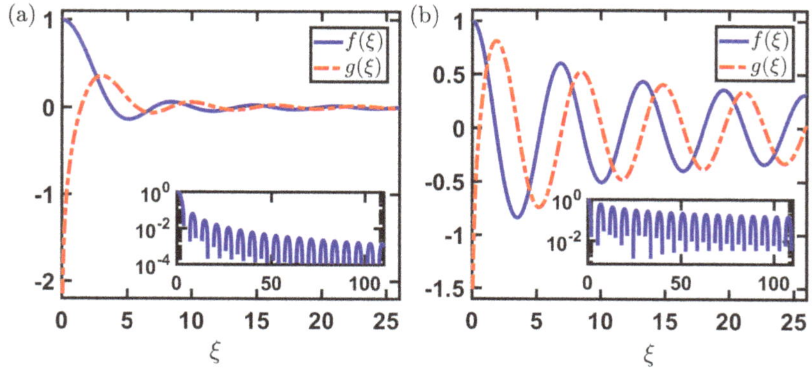

Figure 2.3. Real ($f(\xi)$) and imaginary parts ($g(\xi)$) of 2D dipole–dipole interactions $J_{\mu,\nu}$. The cases for for light polarizations (a) parallel and (b) perpendicular to the atomic separations $\hat{\mathbf{r}}_{\mu,\nu}$ are presented along with respective insets of $f(\xi)$ in longer distances.

trend than the case of 3D dipole–dipole interactions $\sim 1/\xi^3$. We note that the collective frequency shift of $\Omega_{\mu,\nu}$ becomes $-\infty$ for both cases of parallel and perpendicular dipole orientations in figure 2.3, which is in contrast to the 3D case in figure 2.2 where corresponding $\Omega_{\mu,\nu}$ goes to $\mp\infty$, respectively. We attribute this difference to the prefactor of $\hat{\mathbf{p}} \cdot \hat{\mathbf{r}}_{\mu,\nu}$ in $\Omega_{\mu,\nu}$ where parallel and orthogonal dipole orientations change signs of $\Omega_{\mu,\nu}$ in the 3D case as $\xi \to 0$.

For a long range of $\xi \gg 1$, the asymptotics of $f(\xi) \to 1/\xi^{3/2}$ ($1/\sqrt{\xi}$) for light polarization parallel (perpendicular) to the atomic distances, which again is in contrast to the asymptotic form in free space of $1/\xi^2$ ($1/\xi$). We plot this rather long-range dependence of $1/\sqrt{\xi}$ in figure 2.3(b) and its inset, where the dissipation strength can still be kept at $0.1\Gamma_{2D}$ lasting over $\xi \sim 100$. This presents the distinctive feature of 2D functional forms that support atom–atom interactions to rather longer distances than the 3D case. In figure 2.3(a) and its inset for another dipole orientation, the strength also keeps $10^{-3}\Gamma_{2D}$ at $\xi \sim 100$, still in huge contrast to the 3D case in figure 2.2. This will have a significant effect on the properties of collective emissions (super- and subradiance), which we will investigate further in chapter 4.

2.4 One-dimensional dipole–dipole interactions

Here I further reduce one more dimension in space and show the results of 1D dipole–dipole interactions. From equation (2.6), we consider a 1D reservoir which gives a quantization volume of L in $\bar{g}_q$ and obtain

$$J_{\mu,\nu} = \int_{-\infty}^{\infty} \frac{dq}{2\pi} g_q^2 L e^{i\mathbf{k}_q\cdot(\mathbf{r}_\mu - \mathbf{r}_\nu)} \left[\pi\delta(\omega_q - \omega_e) + \pi\delta(\omega_q + \omega_e) + i\mathcal{P}(\omega_e - \omega_q)^{-1} - i\mathcal{P}(\omega_q + \omega_e)^{-1}\right]. \tag{2.23}$$

We further obtain (let $x_{\mu,\nu} = x_\mu - x_\nu$ denote linear spatial coordinates and remove q in ω_q for brevity)

$$J_{\mu,\nu} = \int_0^{\infty} \frac{d\omega}{\pi} |\partial_\omega q(\omega)| g_q^2 L \cos(\mathbf{k}_q x_{\mu,\nu}) [\pi\delta(\omega - \omega_e) + \pi\delta(\omega + \omega_e) + i\mathcal{P}(\omega_e - \omega)^{-1} - i\mathcal{P}(\omega + \omega_e)^{-1}]. \tag{2.24}$$

Let $\Gamma_{1D} \equiv 2|\partial_\omega q(\omega)|_{\omega=\omega_e} \bar{g}_{k_L}^2 L$, where we assume the dispersion relation of the 1D coupling is constant, and we then obtain

$$J_{\mu,\nu} = \frac{\Gamma_{1D}}{2} \cos(k_L x_{\mu,\nu}) - \frac{i\mathcal{P}}{\pi} \int_{-\infty}^{\infty} d\omega \frac{\mathrm{Re}\left[|\partial_\omega q(\omega)| \bar{g}_q^2 L e^{i\mathbf{k}_q (x_\mu - x_\nu)} \right]}{\omega - \omega_e}, \tag{2.25}$$

which shows the Kramers–Kronig relation between real and imaginary parts of $J_{\mu,\nu}$. Finally the cooperative spontaneous emissions in a 1D reservoir becomes [12]

$$J_{\mu,\nu} = \frac{\Gamma_{1D}}{2} \left[\cos(k_L x_{\mu,\nu}) + i \sin(k_L |x_{\mu,\nu}|) \right]. \tag{2.26}$$

The above sinusoidal form shows an infinite-range feature of dipole–dipole interactions between any pair of atoms. This interaction form, in huge contrast to 2D and 3D results presented in previous sections, shows no divergence in collective frequency shifts. Moreover, a genuine Dicke regime of $J_{\mu,\nu} = \Gamma_{1D}/2$ without dispersion at $\xi = 2m\pi$ for integers m ($\sin(\xi)$ becomes vanishing) can be retrieved in this 1D reservoir. This suggests that such 1D system allows a strong interaction regime that higher dimensional systems cannot achieve. Meanwhile, for the parameter regimes of $\xi = \pi/2$ or $3\pi/2$ shown in figure 2.4, such 1D systems are only subject to intrinsic decays individually, and allows coherent exchanges between atoms without collectively enhanced dissipations.

Also plotted in figure 2.4, interesting parameter regimes of $\xi = \pi$ suggest even 'negative' decay rates between the pair of atoms. This seems to be a gain arising from 1D dipole–dipole interactions, but it actually indicates a subradiant regime due to this π-phase coupling and does not violate any physical laws as if entropy decreases or the arrow of time reverses in dissipative systems. It would be evident in the

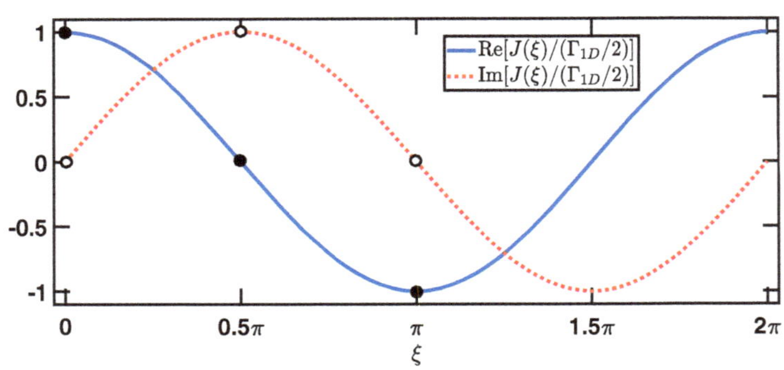

Figure 2.4. Real and imaginary parts of 1D dipole–dipole interactions $J_{\mu,\nu}$. Filled and empty circles represent the interesting parameter regimes, $\xi = 0$, $\pi/2$, or π, with corresponding strengths of decay rates and frequency shifts.

eigenmodes when the individual intrinsic decays are included along with this off-diagonal 'gain'. And one of the eigenmodes in the original Dicke's states is a subradiant state [13], and for two quantum emitters, it is the anti-symmetric state or a singlet state for two spin-1/2 particles $(|\uparrow\downarrow\rangle - |\downarrow\uparrow\rangle)/\sqrt{2}$ that has a π phase difference between $|\uparrow\downarrow\rangle$ and $|\downarrow\uparrow\rangle$. This is the foundation of the subradiant decay channel with a vanishing coupling strength, where in the physics community it is also called the decoherence-free state that does not decay at all.

Like in 2D cases, for 1D systems to show 1D dipole–dipole interactions, we specifically need designed atom–light interfaces. It is only recently that atom–nanofiber, atom–waveguide, or quantum dot–waveguide systems have been designed and experimentally realized to make this long-range dipole–dipole interaction possible. Of course there are always non-radiative losses from heating or surface imperfections of waveguides, but this will not be the bottleneck, which can be overcome by technical improvement, for interesting fundamental studies on emerging phenomena and potential development to strongly interacted quantum interfaces. I will address more of the collective phenomena and potentiality from these 1D systems in Chapter 6.

References

[1] Scully M O and Zubairy M S 1997 *Quantum Optics* (Cambridge: Cambridge University Press)

[2] Lehmberg R H 1970 Radiation from an N-atom system. I. General formalism *Phys. Rev.* A **2** 883–8

[3] Zangwill A 2013 *Modern Electrodynamics* (Cambridge: Cambridge University Press)

[4] Scully M O 2009 Collective Lamb shift in single photon Dicke superradiance *Phys. Rev. Lett.* **102** 143601

[5] Jen H H 2015 Superradiant cascade emissions in an atomic ensemble via four-wave mixing *Ann. Phys.* **360** 556–70

[6] James D F V 1993 Frequency shifts in spontaneous emission from two interacting atoms *Phys. Rev.* A **47** 1336–46

[7] Mittal S, Goldschmidt E A and Hafezi M 2018 A topological source of quantum light *Nature* **561** 502–6

[8] Yaghjian A D 1980 Electric dyadic Green's function in the source region *Proc. IEEE* **68** 248–63

[9] Máximo C E, Piovella N, Courteille P W, Kaiser R and Bachelard R 2015 Spatial and temporal localization of light in two dimensions *Phys. Rev.* A **92** 062702

[10] Longo P, Keitel C H and Evers J 2016 Tailoring superradiance to design artificial quantum systems *Sci. Rep.* **6** 23628

[11] Jen H H 2019 Super- and sub-radiance from two-dimensional resonant dipole-dipole interactions *Sci. Rep.* **9** 5804

[12] González-Tudela A and Porras D 2013 Mesoscopic entanglement induced by spontaneous emission in solid-state quantum optics *Phys. Rev. Lett.* **110** 080502

[13] Dicke R H 1954 Coherence in spontaneous radiation processes *Phys. Rev.* **93** 99–110

IOP Publishing

Collective Light Emission
Many quantum emitters
Hsiang-Hua Jen

Chapter 3

Construction of Hilbert space

Here I introduce a Hilbert space responsible for collective light emissions. In general there are infinite ways to construct a complete Hilbert space, as long as they satisfy the normalization and orthogonality criteria required for a genuine quantum mechanical treatment of the system dynamics. I first review Dicke's space from many spin-1/2 particles and then introduce a couple of other Hilbert spaces with single excitations. Finally a multiply-excited space is discussed, which is useful for studying the collective multiphoton states.

3.1 Dicke–Hilbert space

The superradiant light emission can be understood in terms of Dicke's ladder states with an analogy to spin-1/2 quantum systems [1]. This system can be further characterized by the numbers of up and down spins, n_+ and n_- respectively. And as such, a complete Hilbert space of energy eigenstates can be found and characterized by the total angular momenta j and the magnetic moments $m = (n_+ - n_-)/2$ with a conservation of total number of atoms $N = n_+ + n_-$. These two quantum numbers for the space $|j, m\rangle$ are good and chosen as the common eigenstates for spin operators $\hat{J}_z$ and $\hat{J}^2$ since $[\hat{J}_z, \hat{J}^2] = 0$ [2].

As an example of few-atom systems, I take two and four spin-1/2 particles and demonstrate their energy eigenstates in figure 3.1. The number degeneracy of the states for particular m can be obtained from permutations $N!/[(N/2 + m)!(N/2 - m)!]$. The allowable j is always larger than zero and has a maximum of $N/2$. For a sector of $j = N/2$, it always has a spin exchange symmetry that leads to symmetric states. For other j, only anti-symmetric state of a singlet exists for $N = 2$. For $N \geqslant 3$, $j \neq N/2$ does not involve particular symmetries and thus cannot be determined uniquely [2].

doi:10.1088/978-0-7503-2699-5ch3

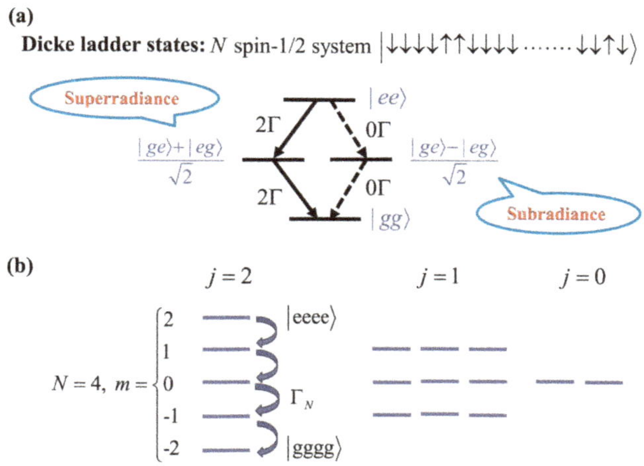

Figure 3.1. Dicke ladder states for (a) $N = 2$ and (b) $N = 4$ as examples. These complete Hilbert space can be expressed in terms of spin-1/2 systems with up (excited state $|e\rangle$) and down (ground state $|g\rangle$) spins. Different eigenstates are characterized by total angular momenta j and magnetic moments m. In particular, super- and sub-radiance are identified for $N = 2$ with triplet and singlet transition channels in (a), and Γ_N in (b) from a half excited state in the sector of $j = 2$ and $m = 0$ decays to $m = -1$, leading to the most significant superradiant emission.

The rate of photo emissions is proportional to $\langle \hat{J}_+ \hat{J}_- \rangle$ where $\hat{J}_\pm = \hat{J}_x \pm i\hat{J}_y$, which leads to

$$\langle j, m \mid \hat{J}_+ \hat{J}_- \mid j, m \rangle \propto (j + m)(j - m + 1). \tag{3.1}$$

This gives a maximal emission rate of $N^2/4$ when $j = N/2$ and $m = 0$ for $N \gg 1$, associated with the half excited states. A significant enhancement of N^2 in the light intensity therefore attributes to superradiance which shows a shortened emission time proportional to the number of the emitters.

Finally in this section, I give a remark on some interesting observations of Dicke states. First, these Dicke states are highly correlated, and for example of the sector of $J = N/2$ with $m \neq \pm N/2$, they are maximally entangled. Therefore, we can be sure that superradiant emissions are highly related to the quantum coherences within the systems. Secondly, the collective dipole moment $\langle \hat{J}_+ \rangle$ under Dicke states vanishes. The superradiance with a high intensity actually originates from a negligible dipole moment, which is a highly nonclassical phenomenon. We note that a product state of N coherent quantum emitters,

$$|\psi\rangle = \left(\frac{|g\rangle + |e\rangle}{\sqrt{2}} \right)^{\otimes N}, \tag{3.2}$$

also presents an enhanced photo emission rate, but with a significant collective dipole moment. This behaves more like the classical dipoles and is called superfluorescence [2]. Therefore, an enhanced light radiation is not sufficient to be classified as superradiance, which needs more information on the properties of the collective dipole moment and quantum correlations.

3.2 Singly-excited Hilbert space

Here we introduce some possible ways to construct a complete Hilbert space with single excitation of light. The subject of single photon excitation is interesting itself, especially in the perspective of single photon source generation and its application in quantum information processing. Aside from this, when an interaction of light and matter is involved, how the atomic system is excited and evolved in time is another topic for continuing investigation.

Naturally for singly-excited atomic states, a bare state basis is a good starting point. This basis is

$$|\psi_m\rangle = |e\rangle_m |g\rangle^{\otimes(N-1)}, \tag{3.3}$$

where $m \in [1, N]$, $\mathbf{k}$ is the wave vector of a single photon, and the tensor product denotes the $(N-1)$ ground state atoms. Here we may add a traveling phase $e^{i\mathbf{k}\cdot\mathbf{r}_m}$ on the excited atom to denote a genuine spatial consideration of an atomic ensemble. This will lead to an overall geometrical factor on the superradiant decay rate when the resonant dipole–dipole interactions are considered in the dissipation of the system. Additionally, this traveling phase is unavoidable when the light propagation effect is significant especially in a long atomic ensemble. The bare state basis, however, is not in general an eigenstate basis for the system dynamics. Below we proceed to construct a more intuitive basis that takes care of the timescale and dynamical evolutions of the system.

When a single photon interacts with the whole ensemble of atoms, we do not know which one is excited with certainty. Therefore, a symmetrical and singly-excited state or timed Dicke state [3] is formed on the absorption of such photon,

$$|\phi_N\rangle = \frac{1}{\sqrt{N}} \sum_{\mu=1}^{N} e^{i\mathbf{k}\cdot\mathbf{r}_\mu} |e\rangle_\mu |g\rangle^{\otimes(N-1)}. \tag{3.4}$$

Since a complete Hilbert space of N atoms with one excitation should have N eigenstates, we expect that other $(N-1)$ states are orthogonal to the symmetrical one. We note that a basis spanned from the symmetrical state is again not an eigenbasis either. This is due to the nature of long-range dipole–dipole interactions discussed in chapter 2.

A complete Hilbert space for a single photon interacting with the atomic ensemble has been proposed [4], and we take it as one example,

$$|\phi_m\rangle = \sum_{\mu=1}^{N} \left(C - \frac{\sqrt{N}+1}{N-1}\delta_{\mu N} - \delta_{\mu m} \right) e^{i\mathbf{k}\cdot\mathbf{r}_\mu} |\psi_\mu\rangle, \tag{3.5}$$

for $m \in [1, N-1]$ with

$$C \equiv \frac{1 + 1/\sqrt{N}}{N - 1}. \tag{3.6}$$

A complete basis requires orthogonality relations and normalization, and the above basis $|\phi_m\rangle$ satisfies these requirements. This construction of Hilbert space actually can lead to many as long as they meet the criteria of a complete Hilbert space. Actually there are infinite ways of constructions, and therefore it makes no sense to create another one, unless the basis by construction can be accessible via manipulations or external controls in practical implementations.

As another example, we show a construction of a complete space in terms of Dicke's states, where singlet and triplet states for $N = 2$ provide the foundation [5],

$$|\phi_m\rangle = \frac{\sum_{\mu=1}^{m} e^{i\mathbf{k}\cdot\mathbf{r}_\mu}|\psi_\mu\rangle - m e^{i\mathbf{k}\cdot\mathbf{r}_{m+1}}|\psi_{m+1}\rangle}{\sqrt{m(m+1)}}, \tag{3.7}$$

for $m \in [1, N-1]$.

These two examples of Hilbert space mark two features on constructing a complete space with single excitation. The first one involves all the atoms, but each involves different probability amplitudes, whereas the second involves only a part of the atoms except the symmetrical state. I will later introduce the third option for a complete basis with uniform probability amplitudes on all the atoms, which benefits from a global control on the constituent atoms, instead of requiring partial or individual dressing fields that will complicate the process of state preparation and engineering.

We note that the above two examples of complete bases reduce to a bare state basis when $N \gg 1$. For equation (3.5), the coefficients in $|\phi_{m\neq N}\rangle$ become the most pronounced in the bare state of $|\psi_m\rangle$ with an order of one, while the others are in the order of N^{-1} or $N^{-1/2}$. For equation (3.7), when $m \sim \mathcal{O}(N)$, we again obtain significant probability amplitudes in the bare state of $|\psi_{m+1}\rangle$ with an order of one, that is $|\phi_m\rangle \sim |\psi_{m+1}\rangle$, while the others are in the order of m^{-1}.

Now I introduce the third option to construct a complete basis, which correlates all the atoms collectively with a uniform probability amplitude. They will never reduce to the bare state basis, showing the nature of the correlation. They are [6, 7]

$$|\phi_m\rangle = \sum_{\mu=1}^{N} \frac{e^{i\mathbf{k}\cdot\mathbf{r}_\mu}}{\sqrt{N}} e^{i\frac{2m\pi}{N}(\mu-1)}|\psi_\mu\rangle, \tag{3.8}$$

where the normalization is satisfied with the orthogonality relations

$$\langle \phi_m | \phi_n \rangle = \frac{1}{N} \sum_{\mu=1}^{N} e^{i\frac{2\pi}{N}(\mu-1)(m-n)} = \delta_{m,n}. \tag{3.9}$$

This relation is supported by De Moivre's formula, which is originally used for the nth root of unity for a complex number z such that $z^n = 1$. The coefficients in the above are obtained by the roots of $z^N = 1$, which are $e^{i2\pi m/N}$ for $m \in [1, N]$. The orthogonality relations are guaranteed by the identity of the sum of all roots, $\mathrm{SR}(n) = 0$ for all n except for $n = 1$. To prepare various De Moivre states as shown in figure 3.2, the phases of the atoms can be imprinted by a pulsed and linearly increased magnetic field gradient in 1D [6] or 3D [7] optical lattices using unitary transformations. Pulsed light fields on the atoms to create AC-Stark shifts can also be the alternative method to imprint the phases on the atoms. We note that the form of De Moivre states in equation (3.8) with $\mathbf{r}_\mu \to 0$ is first constructed in the degenerate subspace of a small sample in Agarwal's book [8].

3.3 Multiply-excited Hilbert space

For a Hilbert space of multi-excitation atomic states, again there are infinite ways to construct such bases. One option is to extend the results of singly-excited space [4], which leads to M-excitation states ($M \leqslant N$)

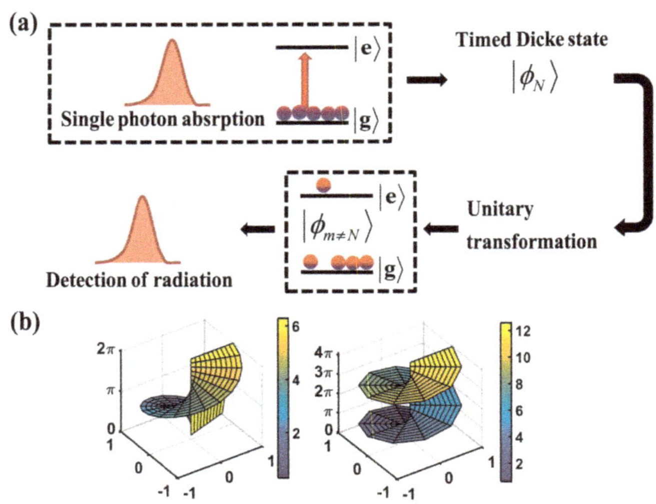

Figure 3.2. Schematic preparation of De moivre states. (a) Various De Moivre states can be prepared by an absorption of single photon, followed by a unitary transformation to $|\phi_{m\neq N}\rangle$ from the timed Dicke state $|\phi_N\rangle$. (b) Examples of incremental phase changes for $m = 1$ (left) and $m = 2$ (right) in equation (3.8) respectively, where $N = 20$ is used for illustration.

$$
M = 0, \quad |\phi_0\rangle = |0\rangle^{\otimes N}.
$$

$$
M = 1, \quad \begin{cases} |\phi_N\rangle = \dfrac{1}{\sqrt{N}} \displaystyle\sum_{\mu=1}^{N} e^{i\mathbf{k}\cdot\mathbf{r}_\mu}|\psi_\mu\rangle; \\ |\phi_m\rangle = \displaystyle\sum_{j=1}^{N-1}\left(\dfrac{1+1/\sqrt{N}}{N-1} - \delta_{jm}\right)e^{i\mathbf{k}\cdot\mathbf{r}_j}|\psi_j\rangle - \dfrac{e^{i\mathbf{k}\cdot\mathbf{r}_N}}{\sqrt{N}}|\psi_N\rangle, \quad m = 1, 2, \dots, N \\ - 1. \end{cases}
$$

$$
M = 2, \quad \begin{cases} |\phi_{N_2}^{(M)}\rangle = \dfrac{1}{\sqrt{N(N-1)/2}} \displaystyle\sum_{\mu>\nu}^{N}\sum_{\nu=1}^{N} e^{i\mathbf{k}\cdot(\mathbf{r}_\mu+\mathbf{r}_\nu)}|\psi_{\mu,\nu}\rangle; \\ |\phi_m^{(M)}\rangle = \displaystyle\sum_{l>j}^{N}\sum_{j=1}^{N-2}\left(\dfrac{1+1/\sqrt{N_2}}{N_2-1} - \delta_{m,(j,l)}\right)e^{i\mathbf{k}\cdot(\mathbf{r}_j+\mathbf{r}_l)}|\psi_{j,l}\rangle \\ -\dfrac{e^{i\mathbf{k}\cdot(\mathbf{r}_{N-1}+\mathbf{r}_N)}}{\sqrt{N_2}}|\psi_{N-1,N}\rangle, \quad m = 1, 2, \dots N_2 - 1, \text{ and } N_2 \equiv N(N-1)/2; \end{cases} \tag{3.10}
$$

$$
\vdots
$$

$$
M = N, \quad |\phi_1^{(N)}\rangle = |e\rangle^{\otimes N}\prod_{j=1}^{N} e^{i\mathbf{k}\cdot\mathbf{r}_j},
$$

where two excitation bare states are

$$
|\psi_{\mu,\nu}\rangle \equiv |e\rangle_\mu |e\rangle_\nu |g\rangle^{\otimes(N-2)}, \tag{3.11}
$$

and (j, l) in the subscript of the Kronecker delta function of two excitation states is defined so that $(1, 2) = 1$, $(1, 3) = 2, (1, 4) = 3, \dots, (1, N) = N - 1$, $(2, 3) = N$, $(2, 4) = N + 1, \dots$, and so on. The upper script on each multi-excitation state denotes the number of excitations, and it is straightforward to construct the states for $M \geqslant 3$ with a normalization of N_M and additional labels of $(j, l, \dots, p)$ in Kronecker delta functions. These states are constructed in the same spirit in forming singly-excited state basis, which are obtained by one symmetric state $|\phi_{N_n}^{(n)}\rangle$ ($N_n = C_n^N$ as binomial coefficients) and $(N_n - 1)$ non-symmetric ones. Note that the ground $|\phi_0\rangle$ and fully excited states $|\phi_{N_n}\rangle$ are symmetric, and the sum of all these number of states is exactly $2^N = \sum_{n=0}^{N} N_n$, reflecting the degrees of freedom in all accessible states provided by N two-level quantum emitters.

Similarly, when $N \gg 1$, these multi-excitation states approach the bare states. Therefore, these states cannot reflect the collective nature of field excitations on the atoms. Here I further introduce another construction of multi-excitation states. We consider a setting where, in general, a resonant M multiphoton is absorbed by N two-level quantum emitters, which leads to a symmetrical and multiply-excited state,

$$
|\phi_{N_M}^{(M)}\rangle = \frac{1}{\sqrt{N_M}} \prod_{j=1}^{M} \sum_{\mu_j=\mu_{j-1}+1}^{N-M+j} e^{i\mathbf{k}\cdot\sum_{m=1}^{M}\mathbf{r}_{\mu_m}} |e\rangle_{\mu_1}|e\rangle_{\mu_2}\cdots|e\rangle_{\mu_M}|g\rangle^{\otimes(N-M)}, \tag{3.12}
$$

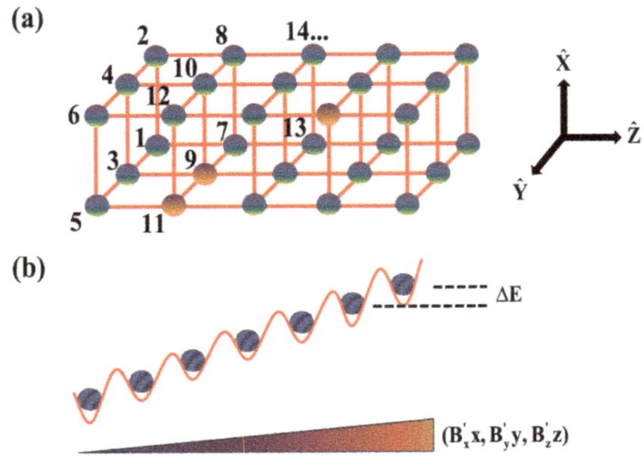

Figure 3.3. Labeling and phase imprinting in a 3D atomic array. The labels in (a) denote the specific ordering of linearly increasing phases, which can be imprinted by three Zeeman magnetic gradient fields (B'_x, B'_y, B'_z) in (b). A unitary transformation of $\hat{T} = e^{i\Delta E\tau_B}$ on the atoms evolves the timed Dicke state toward one of the multi-excitation states, where $\Delta E(x, y, z) = B'_x x + B'_y y + B'_z z$ multiplied by an interaction time τ_B denotes the amount of phases imprinted on the array from linearly increasing energy spacings in (b) schematically.

where $\mu_0 = 0$, and $\mathbf{k} \cdot \sum_{m=1}^{M} \mathbf{r}_{\mu_m} \equiv \mathbf{k} \cdot \mathbf{R}_M$ denotes the traveling phase on the absorption of a multiphoton source. The M atomic excitations are distributed symmetrically among N atoms with N_M possible configurations. For a concise notation, we define the bare states as $|\psi_N^{(M)}(\vec{\mu})\rangle \equiv |e\rangle_{\mu_1}|e\rangle_{\mu_2}\cdots|e\rangle_{\mu_M}|g\rangle^{\otimes(N-M)}$ with $\vec{\mu} \equiv (\mu_1, \mu_2, \cdots, \mu_M)$. For an equidistant 3D atomic array, we obtain the phase-imprinted state [9]

$$|\phi_n^{(M)}\rangle = \prod_{j=1}^{M} \sum_{\mu_j=\mu_{j-1}+1}^{N-M+j} \frac{e^{i\mathbf{k}\cdot\mathbf{R}_M}}{\sqrt{N_M}} e^{i\frac{2n\pi}{N_M}[f(\vec{\mu})-1]} |\psi_N^{(M)}(\vec{\mu})\rangle, \tag{3.13}$$

where $n \in [1, N_M]$. $f(\vec{\mu}) = \sum_{i=1}^{M} \mu_i$ represents the incremental and linear increases of the imprinted phases as shown in figure 3.3(a). These μ also need to be labeled linearly in a specific order along these three axes, and the imprinted phases can be determined by a unitary transformation $\hat{T} = e^{i\Delta E\tau_B}$ where the energy shifts ΔE of the atoms can be controlled by the Zeeman magnetic gradient fields (B'_x, B'_y, B'_z) in figure 3.3(b). The above states $|\phi_n^{(M)}\rangle$, in general, do not form complete and orthogonal states in a multiphoton Hilbert space of M excitations, except $M = 1$ in single-excited De Moivre states. Nevertheless, the states in equation (3.13) can be prepared simply by applying external field gradients, which can therefore be enabled to be manipulated and engineered.

References

[1] Dicke R H 1954 Coherence in spontaneous radiation processes *Phys. Rev.* **93** 99–110

[2] Mandel L and Wolf E 1995 *Optical Coherence and Quantum Optics* (Cambridge: Cambridge University Press)

[3] Scully M O, Fry E S, Ooi C H R and Wódkiewicz K 2006 Directed spontaneous emission from an extended ensemble of N atoms: timing is everything *Phys. Rev. Lett.* **96** 010501
[4] Mazets I E and Kurizki G 2007 Multiatom cooperative emission following single-photon absorption: Dicke-state dynamics *J. Phys. B: At. Mol. Opt. Phys.* **40** F105–112
[5] Svidzinsky A A, Chang J T and Scully M O 2008 Dynamical evolution of correlated spontaneous emission of a single photon from a uniformly excited cloud of N atoms *Phys. Rev. Lett.* **100** 160504
[6] Jen H H, Chang M S and Chen Y C 2016 Cooperative single-photon subradiant states *Phys. Rev.* A **94** 013803
[7] Jen H H 2016 Cooperative single-photon subradiant states in a three-dimensional atomic arra *Ann. Phys.* **374** 27–34
[8] Agarwal G S 1974 Quantum statistical theories of spontaneous emission and their relation to other approaches *Springer Tracts in Modern Physics: Quantum Optics* vol 70 (Berlin: Springer)
[9] Jen H H 2017 Phase-imprinted multiphoton subradiant states *Phys. Rev.* A **96** 023814

IOP Publishing

Collective Light Emission
Many quantum emitters
Hsiang-Hua Jen

Chapter 4

Cooperative spontaneous emissions: super- and subradiance

The phenomenon of cooperative spontaneous emissions has been studied extensively and intensively over the years. Their properties are mainly dependent on the interatomic distances, macroscopic geometry, the number of emitters, and how many photons involved. I first give an update for the recent progress on the subject of collective emissions and then summarize some common approaches applied to the subject. Particular one-dimensional and three-dimensional atomic arrays are investigated, and the cooperative radiation from multiphoton excitations is discussed at the end.

4.1 Recent progress

Strong and collective light–matter interactions in quantum optical systems have enabled tunable and efficient operations in generating, storing, and manipulating quantum correlations [1]. Quantum correlations in neutral atoms are often established by exchanging spontaneous emissions. The spontaneous emissions from an optically thick atomic system are superradiant [2, 3] due to the resonant and pairwise dipole–dipole interactions [4] of the atoms induced in the dissipation process. This collective light–matter interaction also initiates the associated frequency shifts [5–15], and is responsible for subradiant radiations [16] as an afterglow of superradiance [17].

Recent developments involve collective light scattering in a dense atomic medium [18–20] and atomic rings [21, 22], subradiant state preparations [23–28], selective radiance [29] by engineering atomic positions, collective antiresonances from the subradiant arrays in a cavity [30], subradiant excitations in arrays of atoms [31] or metamolecules [32], and quantum optical analogs of topological systems in two-dimensional atomic arrays [33].

doi:10.1088/978-0-7503-2699-5ch4

Most interestingly, a universal scaling of N^{-3} in the most subradiant modes can be obtained in a two-level atom chain coupled to a one-dimensional waveguide [34], which shows fermionic features of spatial repelling in multiply-excited subradiant states [34, 35]. Subradiant or dark states can also be prepared in distant atomic arrays [36] or fermionic atoms in doubly filled optical lattices [37]. Finally, subradiant excitation transport [38] can be realized in a finite linear atomic chain or ring. These provide wide-range applications in quantum state engineering, quantum transport, and quantum communication, and fruitful opportunities in studying many-body physics of light–matter interactions.

Equipped with the flexibility of manipulating the atomic positions utilizing a photonic crystal waveguide [39], optical microtraps [40, 41], or a diamond nano-photonics platform [42], arbitrary spatial distributions of the atoms can be designed and fabricated, beyond the diffraction limit of wavelength of the excitation field. Therefore, this offers many opportunities to explore the super- and subradiant modes, and opens up a new avenue in tailoring and modifying the quantum states of light or matter.

4.2 Methodology

There are many theoretical approaches to study the cooperative spontaneous emissions from many quantum emitters. We list some of them that have their own flavors and angles to investigate such collective phenomena, while each has its own merits or demerits in solving the problems involving light–matter interactions. Nevertheless, these approaches complement each other in unraveling the mechanism responsible for the collective emissions.

The first is the QED (quantum electrodynamics) approach [43], which utilizes photon and atom propagation operators to obtain dipole–dipole scattering amplitudes and collective Lamb shifts for resonant atoms. The usefulness of the diagrammatic techniques leads to the derivations of spectral functions and various Green's functions, which are essential to the collective effect of resonant scatterings between many atoms.

The second approach is the quantum correction method [44], where a mean-field treatment is applied to include the effect of surrounding atoms on the effective probe atom by using the Lorentz–Lorenz equation. This gives an estimate of the collective emission rates for totally inverted systems, while an extension to go beyond the mean field by including higher-order corrections or essentially multiple scattering events is difficult. The next approach takes the stochastic nature of spontaneous emissions and is called the quantum trajectory method [45]. The quantum jump operators are used to demonstrate the collective phenomenon of radiations, which gives an insight into its initiations and allows numerical simulations for a finite number of atoms.

The fourth method is the Schrödinger equations, where the system is evolved within a chosen Hilbert space, and system dynamics can be solved by eigen-decompositions. This is particularly useful for low excitations, but the complexity of this method rapidly grows as the Hilbert space increases for a large number of atoms and excitations. Another method of semi-classical treatment of the Maxwell–Bloch equation is operational and again useful to describe the superradiant light propagation through an atomic medium. Lastly, a positive-P phase-space representation method [46] can be used to include the quantum fluctuations, which are origins of emerging superradiance from a dense atomic gas. This shows the essence of quantum noises that are indispensable to the collective radiations.

I only present several theoretical approaches that have been popular or reinvigorated in their time. These approaches take the roles of either atom or photon, or both to show the collectiveness of light–matter interacting systems. The cooperative spontaneous emissions are therefore emerging phenomena related to quantum fluctuations from environment[1], with randomness in emissions and with influences from quantum emitters all around. This multiple scattering process in exponentially growing Hilbert space of quantum systems presents one of the delicate and intricate dynamical systems with unlimited complexity. However, we can always go around the apparent obstacle in theoretical formalisms and find a way to reveal and discover essential physics of interests relevant to experimental observations.

As a final remark, a stark contrast between noninteracting and highly interacting systems that initiate cooperative spontaneous emissions lies not only in the quantity of decay rates, but also the number of them. An effective change of the decay rate is relatively straightforward, either enhanced or reduced, reflecting the role of dipole–dipole interactions that, respectively, cause the constructive or destructive interferences of light radiations [29]. Unfortunately, a single quantity of effective decay rate cannot fully describe what is really happening in highly dynamically coupled light–matter interacting systems. This further leads to a concern of the validity of mean-field Maxwell–Bloch equation treatment [48, 49]. What is more appropriate is to consider a multiple scattering process due to light-induced dipole–dipole interactions, which presents a modification in the Beer–Lambert law [48] in describing the absorption and determining the optical depth of a system, especially in an optically thin 2D gas. Therefore, a description of including multiple decay rates or essentially the multi-modes in radiations should be more appropriate in investigating the cooperative spontaneous emissions.

4.3 One-dimensional atomic array

A one-dimensional (1D) atomic array demonstrates one particular physical system in many other 1D platforms, for example a strongly correlated matter of the Luttinger liquid [50]. This 1D atomic array can be realized in optical lattices of neutral atoms, ion chain, superconducting qubits, quantum dot arrays, or optical microtraps of atoms. This system manifests one of the simplest design of matter in a low dimension of space, while still endowing strong correlations between each quantum emitters via light–matter interactions.

For a 1D atomic array in free space, which experiences 3D resonant dipole–dipole interactions equation (2.6), an eigen-decomposition can be applied to diagonalize

[1] A bath of bosonic particles that can exchange energy to and from the system. Sometimes a 'reservoir' can be used in parallel to an environment to show the large degrees of freedom and faster timescale for equilibrium between system–reservoir interactions, a validity for the Markov approximation [47].

equation (2.6) under singly-excited state Hilbert space $|\psi_\mu\rangle$ in equation (3.3). The state of the atomic system in the Schrödinger picture can be expressed as $|\Psi(t)\rangle = \sum_{\mu=1}^{N} c_\mu(t)|\psi_\mu\rangle$, where the time evolutions of probability amplitudes $c_\mu(t)$ are [24, 25]

$$\dot{c}_\mu(t) = \sum_{\nu=1}^{N} J_{\mu\nu} c_\nu(t). \tag{4.1}$$

We then solve the above coupled equations by a similarity transformation $\hat{U}$ in diagonalizing $J_{\mu\nu}$ along with the eigenvalues λ_n,

$$c_\mu(t) = \sum_{\nu,n} U_{\mu n} e^{\lambda_n t} U_{n\nu}^{-1} c_\nu(t=0), \tag{4.2}$$

where $c_\nu(t=0)$ is an initial probability amplitude of the system.

Since DM states of equation (3.8) are good candidates for a controllable Hilbert space via external manipulations by imprinting spatially-varying phases, we assume that the system is initially in one of the DM states $|\phi_m\rangle$. The evolving state vector $|\Psi(t)\rangle$ is then represented by $\sum_{m'=1}^{N} d_{m'}(t)|\phi_{m'}\rangle$ where $d_{m'}(t=0) = \delta_{m'm}$. From the coefficient transformation between the bare and DM states,

$$d_m = \frac{1}{\sqrt{N}} \sum_{\mu=1}^{N} c_\mu e^{-i\mathbf{k}\cdot\mathbf{r}_\mu} e^{-i\frac{2m\pi}{N}(\mu-1)}, \tag{4.3}$$

we obtain the time evolutions of DM states [24],

$$d_m(t) = \sum_{n=1}^{N} v_n(m) e^{\lambda_n t} w_n(m), \tag{4.4}$$

where

$$v_n(m) \equiv \sum_{\mu=1}^{N} \frac{e^{-i\mathbf{k}\cdot\mathbf{r}_\mu - i2m\pi(\mu-1)/N}}{\sqrt{N}} U_{\mu n}, \tag{4.5}$$

$$w_n(m) \equiv \sum_{\nu=1}^{N} U_{n\nu}^{-1} \frac{e^{i\mathbf{k}\cdot\mathbf{r}_\nu + i2m\pi(\nu-1)/N}}{\sqrt{N}}. \tag{4.6}$$

$\hat{U}$ is composed of the eigenvectors, and $v_n(m)$ is the inner product between the mth DM state and nth eigenstate.

In figure 4.1, we show an example of the 1D atomic array with equal interparticle distances d. Single photon excitation is chosen along $\hat{z}$ with $\hat{x}$ polarization. A significant collective effect emerges when $d/\lambda \ll 1$, which reflects on the subradiant sectors of the ascended eigen-decay rates. When more atoms are involved, smaller subradiant decay rates shows up as well. We can see that the most subradiant eigen-decay rate can be as small as a thousand-fold decrease of the intrinsic one (Γ), which

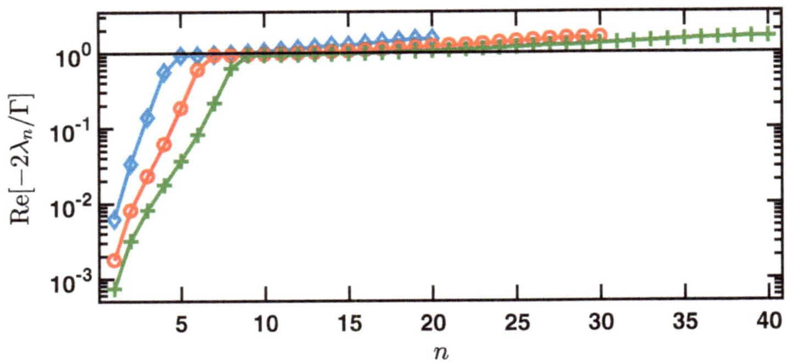

Figure 4.1. Decay rates for a 1D atomic chain. Real parts of the eigenvalues are shown for various $N = 20$ (◇), 30 (○), and 40 (+). Equal inter-particle distance is $d/\lambda = 0.4$ and a horizontal line indicates an intrinsic decay rate Γ.

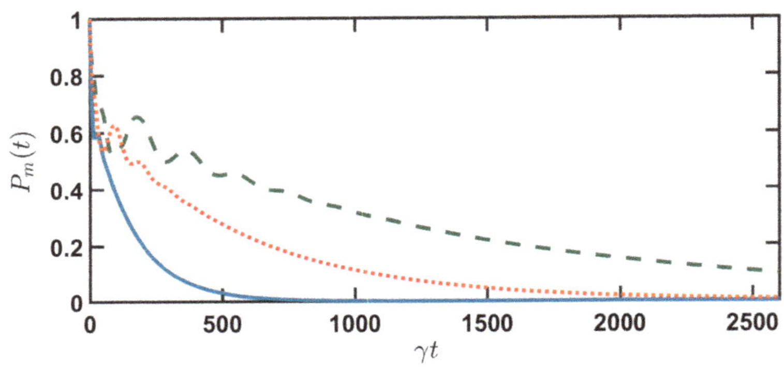

Figure 4.2. Time evolutions of probability $P_m(t)$ for $N = 20$ (solid), 30 (dotted), and 40 (dashed). Respectively the most subradiant DM states are $m = 2$, 3, and 4. It is apparent that $P_4(t)$ at $N = 40$ evolves past $\gamma t \sim 1000$, showing a thousand-fold increase of lifetime. d is the same as in figure 4.1.

is around 26 ns for Rubidium atoms. Therefore, only several tens of atoms have already pushed the system into a lifetime of the microsecond regime. This shows the power of collective RDDI that allows an unprecedented scale of lifetime, which could be a novel resource for quantum information applications or quantum storage in particular. On the other hand, for the superradiant decay rates, there are several superradiant modes that emerge above Γ. For $d/\lambda \ll 1$, the most superradiant decay rate ($n = N$) is enhanced, showing a relatively well-known phenomenon in the early development of collective radiations. As $d/\lambda \gg 1$, RDDI become less significant due to $1/d^3$ dependence, which leads to the independent atom regime where single particle physics with one characteristic scale of Γ in describing the system dissipations is adequate.

In figure 4.2, we locate the maximal $|v_1(m)w_1(m)|^2$ for some m and plot $P_m(t) \equiv |d_m(t)|^2$ as the time evolutions of the most subradiant DM states. As N

increases, longer time dynamics are supported by the system. Some early oscillatory features of $P_m(t)$ present the multimode characteristics that involve different frequency shifts[2] in λ_n, resulting in beatings of time evolutions. This also shows a non-exponential feature that is beyond the independent atom limit.

The results presented here are under perfect conditions, with fixed inter-particle distances and straight line geometry that is parallel to light excitation. Construction of the 1D atomic array may suffer from position fluctuations and misalignment, which can modify RDDI that mediates the whole chain. This compromises the phase coherences between each quantum emitters, and therefore subradiant modes cannot be sustained perfectly. Solid-state qubits (superconducting qubits or quantum dots) can be the potential candidates to surpass the performance of neutral atoms on controlling the subradiant modes, while they themselves also suffer from other issues like non-identical qubit preparations, background noises, and cross talks, which prohibits making a large-scale ensemble of them. On the other hand, a strong coupling regime is relatively easy to be reached for such systems via an external cavity and the guided transmissions.

4.4 Three-dimensional atomic array

Here I further introduce a 3D atomic array, and show how RDDI can modify the eigenmodes and cooperative radiations from a 3D geometry of matter. The preparation of this system has been progressed into arbitrary manipulations of structures [51] and coherent addressing [52, 53] of individual qubits. Therefore, it would be interesting to investigate the properties of super- and subradiant emissions, which can be manipulated by constructing various structures.

In figure 4.3, a comparison between 1D, 2D, and 3D atomic arrays is demonstrated. Almost the same total number of atoms N are used as an example (1D and 3D cases for $N = 125$ and 2D case for $N = 121$). As the dimensions of atoms increase, we can see that the most subradiant mode has lower values for lower dimensional systems, in contrast to the most superradiant modes which is more significant for 3D atomic array. An even lower subradiant mode is presented in figure 4.3(b) for a smaller d. Furthermore, strong RDDI from a smaller d induce more subradiant modes (around 60 below Γ in 1D case) comparing a longer d (around 25 in 1D case).

Next, we present the time evolutions of a DM state that occupies most of the lowest subradiant mode in figure 4.4. Clearly from the normalized weightings on the eigenmodes, $|v_n(m)w_n(m)|^2$, the DM state with $m = 12$ overlaps mostly the first two lowest subradiant modes, which leads to a beating evolution in its probability. More structures of 3D atomic arrays can be found in [25], where under an optically thick array, a case of $d/\lambda \sim 0.6$ can also support significant subradiant evolutions, while more multiple beating structures can be shown in cubic arrays.

[2] Under some conditions of d (usually smaller) and N, we can extract the difference of two main frequency shifts from the beating frequency. This allows a fundamental study on the energy shifts of those eigenmodes in a systematic way under well-controlled system parameters.

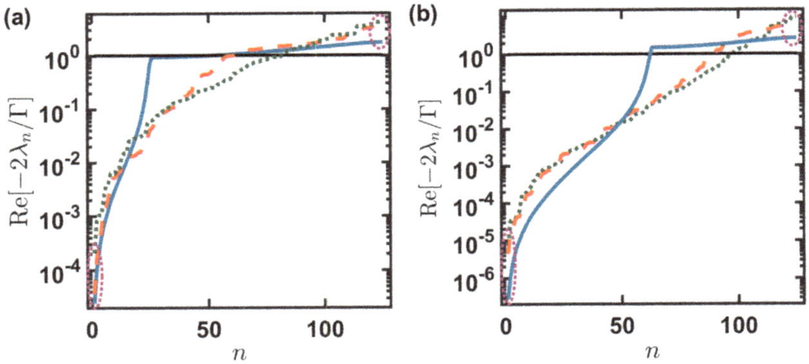

Figure 4.3. Decay rates for a 3D atomic chain. Real parts of the eigenvalues are shown for atoms in various dimensions, 1D chain of $N_z = 125$ (solid), 2D array of $N_x = N_y = 11$ (dashed), and 3D array of $N_x = N_y = N_z = 5$ (dotted). Equal inter-particle distances are chosen as (a) $d/\lambda = 0.4$ and (b) $d/\lambda = 0.25$ for comparisons. A horizontal line indicates an intrinsic decay rate Γ, and dashed circles mark the most sub- and superradiant eigenmodes.

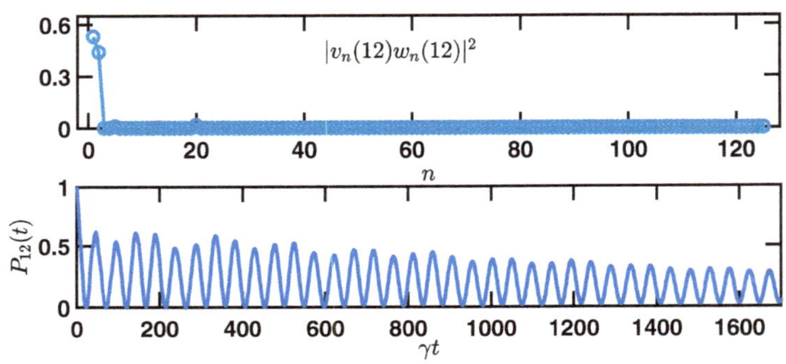

Figure 4.4. Weightings on the eigenmodes and time evolutions of probability $P_m(t)$. For 3D atomic array of $N_x = N_y = N_z = 5$, significant weightings on the most subradiant modes goes to $m = 12$ (○), and its time evolution lasts to $\gamma t \sim 1500$. $d/\lambda = 0.4$.

As a comparison, a 2D case of atomic array under 3D RDDI is plotted in figure 4.5. A longer time dynamic is expected since the most subradiant mode has lower decay rates than a 3D case for the same N. Again similar beating patterns show up due to the frequency shifts associated with the eigenmodes that are mostly occupied by the particular DM state.

4.5 Multiphoton cooperative spontaneous emissions

In this last section of cooperative radiations, I introduce a multiphoton perspective of DM states [28]. This is both experimentally and theoretically of interest since multiphoton interaction with matter is one of the central studies in quantum information processing. Due to the complexity of quantum systems, for M atomic excitations out of a total N atoms ($N \geqslant M$), the allowable quantum states can be up to $\mathcal{O}(N^M)$ when $N \gg M$. The number of accessible states can be quickly exceeding

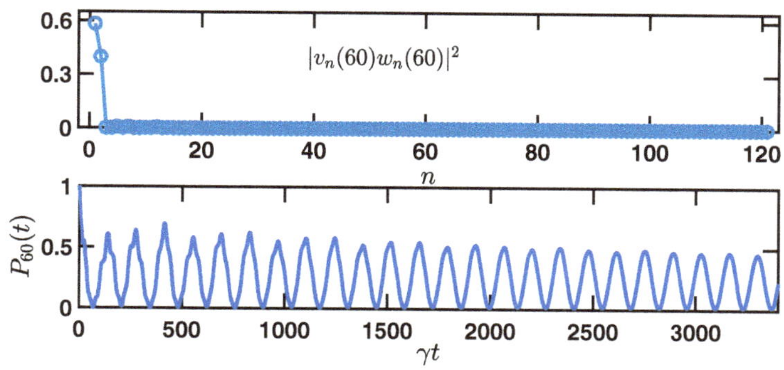

Figure 4.5. Weightings on the eigenmodes and time evolutions of probability $P_m(t)$. For the 2D atomic array of $N_x = N_y = 11$, significant weightings on the most subradiant modes of $m = 60$ (○) lead to its time evolution over $\gamma t \sim 3000$. $d/\lambda = 0.4$.

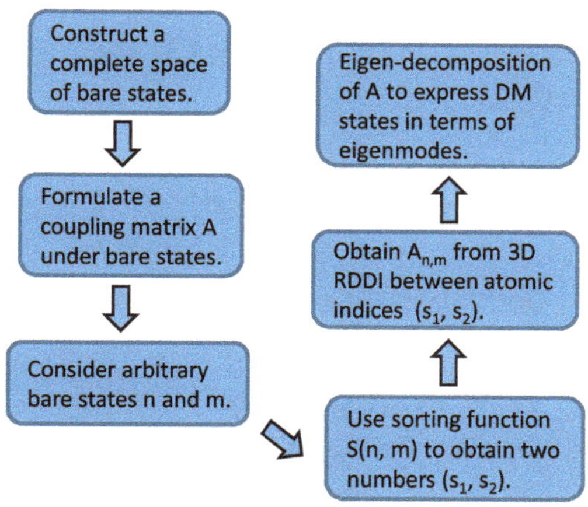

Figure 4.6. Schematic flow charts for constructing a multiphoton coupling matrix. A multiphoton coupling matrix requires a sorting algorithm to obtain the essential coupling constants between every atoms under a bare state basis.

the handling power of classical computers, which presents the supremacy of quantum systems useful for future applications in tackling particular formidable tasks. To have a sense of how an large amount it could be, I take $N = 1000$ with $M = 100$, which gives $\mathcal{O}(10^{300})$. This is unimaginable for such moderate N and M, but in a quantum world, it is common to have such large degrees of freedom that systems can reside, let alone other dimensions of the system can join to create an even larger space in quantum perspective, say multi-levels (hyperfine or Zeeman levels) or quantized motions of atoms.

Since RDDI are pairwise interactions, we need a sorting algorithm to obtain the essential coupling matrix under the bare state basis, which I summarize in figure 4.6. Firstly, I demonstrate an example of bare state basis for $N = 5$ with $M = 2$. A few bare states are shown below out of a total $C_M^N = 10$ of them,

$$\begin{aligned}&|1\rangle = |e_1e_2g_3g_4g_5\rangle, \quad |2\rangle = |e_1g_2e_3g_4g_5\rangle, \quad |3\rangle = |e_1g_2g_3e_4g_5\rangle, \quad |4\rangle = |e_1g_2g_3g_4e_5\rangle, \\ &|5\rangle = |g_1e_2e_3g_4g_5\rangle, \quad |6\rangle = |g_1e_2g_3e_4g_5\rangle, \cdots, \quad |10\rangle = |g_1g_2g_3e_4e_5\rangle.\end{aligned} \tag{4.7}$$

Assuming the atoms are placed in order as $\mu_1 < \mu_2 < \cdots < \mu_N$, the philosophy of the above ordering of bare states is sequentially moving the last excited atom to the last one of the chain, which is followed by moving the next to the last excited state one place to the right together with the last excited one and then sequentially moving the last excited atom to the last one of the chain, until all excited atoms move to the end of the chain. This ordering also applies to 2D or 3D atomic arrays, where DM states from phase imprinting can be constructed from the bare ones.

The multiply-excited states can be expressed as

$$|\Psi(t)\rangle = \sum_{n=1}^{C_M^N} c_n e^{i\mathbf{k}\cdot R_M}|n\rangle, \tag{4.8}$$

where the traveling phase is defined in section 3.3. The coupled equations within multiply-excited states then read,

$$\dot{c}_n(t) = \sum_{m=1}^{C_M^N} A_{nm} c_m(t). \tag{4.9}$$

The coupling matrix A can be obtained by principles in figure 4.6. We use a sorting function $S(n, m)$ to extract two atomic indices which are coupled and exchanged by 3D RDDI. As an example, for $N = 5$ and $M = 2$ again, $S(1, 2) \rightarrow (2, 3)$, $S(2, 6) \rightarrow$ null, and $S(1, 5) \rightarrow (1, 3)$. The pairwise RDDI $J_{\mu,\nu}$ of equation (2.6) couples every other two atoms, and therefore the functionality of S is to sort out two numbers $(s_1, s_2) \in [\mu_1, \mu_M]$ (within the excited atomic labels), where $s_1 \neq s_2$ under the condition that all other excited labels are overlapped. The outcome of 'null' from S means a vanishing coupling since no RDDI have an effect on these bare states.

The time evolutions of the phase-imprinted multiphoton DM state can be seen in more detail in [28], where its eigenmode analysis is similar to the previous sections. Although the intrinsic decay rates for multiply-excited bare states are $M\Gamma$, which is proportional to how many excitations M, we can still locate some parameter regimes that support a lower subradiant eigenmode than the singly-excited one. Since the collective radiation depends on the interparticle distances, the number of atoms, and the geometries of atomic arrays, this presents a rich interplay between these parameters to show many distinct radiation properties.

As a final remark, we note of a recent work on multi-excitation subradiant states [54], which points out a 'twilight' state that superradiantly emits the first photon, while subradiantly emitting the second one. This presents a long-lived photon–photon correlation from the twilight states and again shows the rich opportunities to explore many other cooperative multiphoton phenomena useful for quantum information processing or state engineering.

References

[1] Hammerer K, Sørensen A S and Polzik E S 2010 Quantum interface between light and atomic ensembles *Rev. Mod. Phys.* **82** 1041–93

[2] Dicke R H 1954 Coherence in spontaneous radiation processes *Phys. Rev.* **93** 99–110

[3] Gross M and Haroche S 1982 Superradiance: An essay on the theory of collective spontaneous emission *Phys. Rep.* **93** 301–96

[4] Lehmberg R H 1970 Radiation from an N-atom system. I. General formalism *Phys. Rev.* A **2** 883–8

[5] Friedberg R, Hartmann S R and Manassah J T 1973 Frequency shifts in emission and absorption by resonant systems of two-level atoms *Phys. Rep.* **7** 101–79

[6] Scully M O 2009 Collective Lamb shift in single photon Dicke superradiance *Phys. Rev. Lett.* **102** 143601

[7] Röhlsberger R, Schlage K, Sahoo B, Couet S and Rüffer R 2010 Collective Lamb shift in single-photon superradiance *Science* **328** 1248–51

[8] Keaveney J, Sargsyan A, Krohn U, Hughes I G, Sarkisyan D and Adams C S 2012 Cooperative Lamb shift in an atomic vapor layer of nanometer thickness *Phys. Rev. Lett.* **108** 173601

[9] Meir Z, Schwartz O, Shahmoon E, Oron D and Ozeri R 2014 Cooperative Lamb shift in a mesoscopic atomic array *Phys. Rev. Lett.* **113** 193002

[10] Pellegrino J *et al* 2014 Observation of suppression of light scattering induced by dipole-dipole interactions in a cold-atom ensemble *Phys. Rev. Lett.* **113** 133602

[11] Jen H H 2015 Superradiant cascade emissions in an atomic ensemble via four-wave mixing *Ann. Phys.* **360** 556–70

[12] Jennewein S *et al* 2016 Coherent scattering of near-resonant light by a dense microscopic cold atomic cloud *Phys. Rev. Lett.* **116** 233601

[13] Jenkins S D *et al* 2016 Optical resonance shifts in the fluorescence of thermal and cold atomic gases *Phys. Rev. Lett.* **116** 183601

[14] Peyrot T *et al* 2018 Collective Lamb shift of a nanoscale atomic vapor layer within a sapphire cavity *Phys. Rev. Lett.* **120** 243401

[15] Wen P Y *et al* 2019 Large collective Lamb shift of two distant superconducting artificial atoms *Phys. Rev. Lett.* **123** 233602

[16] Guerin W, Araújo M O and Kaiser R 2016 Subradiance in a large cloud of cold atoms *Phys. Rev. Lett.* **116** 083601

[17] Mazets I E and Kurizki G 2007 Multiatom cooperative emission following single-photon absorption: Dicke-state dynamics *J. Phys. B: At. Mol. Opt. Phys.* **40** F105–12

[18] Bromley S L *et al* 2016 Collective atomic scattering and motional effects in a dense coherent medium *Nat. Commun.* **7** 11039

[19] Zhu B, Cooper J, Ye J and Rey A M 2016 Light scattering from dense cold atomic media *Phys. Rev.* A **94** 023612

[20] Shahmoon E, Wild D S, Lukin M D and Yelin S F 2017 Cooperative resonances in light scattering from two-dimensional atomic arrays *Phys. Rev. Lett.* **118** 113601

[21] Jen H H, Chang M S and Chen Y C 2018 Cooperative light scattering from helical-phase-imprinted atomic rings *Sci. Rep.* **8** 9570

[22] Jen H H 2018 Directional subradiance from helical-phase-imprinted multiphoton states *Sci. Rep.* **8** 7163

[23] Scully M O 2015 Single photon subradiance: Quantum control of spontaneous emission and ultrafast readout *Phys. Rev. Lett.* **115** 243602
[24] Jen H H, Chang M S and Chen Y C 2016 Cooperative single-photon subradiant states *Phys. Rev.* A **94** 013803
[25] Jen H H 2016 Cooperative single-photon subradiant states in a three-dimensional atomic array *Ann. Phys.* **374** 27–34
[26] Sutherland R T and Robicheaux F 2016 Collective dipole-dipole interactions in an atomic array *Phys. Rev.* A **94** 013847
[27] Bettles R J, Gardiner S A and Adams C S 2016 Cooperative eigenmodes and scattering in one-dimensional atomic arrays *Phys. Rev.* A **94** 043844
[28] Jen H H 2017 Phase-imprinted multiphoton subradiant states *Phys. Rev.* A **96** 023814
[29] Asenjo-Garcia A, Moreno-Cardoner M, Albrecht A, Kimble H J and Chang D E 2017 Exponential improvement in photon storage fidelities using subradiance and 'selective radiance' in atomic arrays *Phys. Rev.* X **7** 031024
[30] Plankensteiner D, Sommer C, Ritsch H and Genes C 2017 Cavity antiresonance spectroscopy of dipole coupled subradiant arrays *Phys. Rev. Lett.* **119** 093601
[31] Facchinetti G, Jenkins S D and Ruostekoski J 2016 Storing light with subradiant correlations in arrays of atoms *Phys. Rev. Lett.* **117** 243601
[32] Jenkins S D *et al* 2016 Optical resonance shifts in the fluorescence of thermal and cold atomic gases *Phys. Rev. Lett.* **116** 183601
[33] Perczel J *et al* 2017 Topological quantum optics in two-dimensional atomic arrays *Phys. Rev. Lett.* **119** 023603
[34] Albrecht A, Henriet L, Asenjo-Garcia A, Dieterle P B, Painter O and Chang D E 2019 Subradiant states of quantum bits coupled to a one-dimensional waveguide *New J. Phys.* **21** 025003
[35] Zhang Y X and Mølmer K 2019 Theory of subradiant states of a one-dimensional two-level atom chain *Phys. Rev. Lett.* **122** 203605
[36] Guimond P O, Grankin A, Vasilyev D V, Vermersch B and Zoller P 2019 Subradiant Bell states in distant atomic arrays *Phys. Rev. Lett.* **122** 093601
[37] Piñeiro Orioli A and Rey A M 2019 Dark states of multilevel Fermionic atoms in doubly filled optical lattices *Phys. Rev. Lett.* **123** 223601
[38] Needham J A, Lesanovsky I and Olmos B 2019 Subradiance-protected excitation transport *New J. Phys.* **21** 073061
[39] Goban A *et al* 2015 Superradiance for atoms trapped along a photonic crystal waveguide *Phys. Rev. Lett.* **115** 063601
[40] Barredo D, de Léséleuc S, Lienhard V, Lahaye T and Browaeys A 2016 An atom-by-atom assembler of defect-free arbitrary two-dimensional atomic arrays *Science* **354** 1021–3
[41] Endres M *et al* 2016 Atom-by-atom assembly of defect-free one-dimensional cold atom arrays *Science* **354** 1024–7
[42] Sipahigil A *et al* 2016 An integrated diamond nanophotonics platform for quantum-optical networks *Science* **354** 847–50
[43] Manassah J T 1983 Statistical quantum electrodynamics of resonant atoms *Phys. Rep.* **101** 359–427
[44] Fleischhauer M and Yelin S F 1999 Radiative atom-atom interactions in optically dense media: Quantum corrections to the Lorentz-Lorenz formula *Phys. Rev.* A **59** 2427–41

[45] Carmichael H J and Kim K 2000 A quantum trajectory unraveling of the superradiance master equation *Opt. Commun.* **179** 417–27
[46] Jen H H 2012 Positive-P phase-space-method simulation of superradiant emission from a cascade atomic ensemble *Phys. Rev.* A **85** 013835
[47] Cohen-Tannoudji C, Dupont-Roc J and Grynberg G 1992 *Atom-Photon Interactions: Basic Processes and Applications* (New York: Wiley)
[48] Chomaz L, Corman L, Yefsah T, Desbuquois R and Dalibard J Absorption imaging of a quasi-two-dimensional gas: a multiple scattering analysis *New J. Phys.* **14** 055001
[49] Corman L *et al* Transmission of near-resonant light through a dense slab of cold atoms *Phys. Rev.* A **96** 053629
[50] Giamarchi T 2004 *Quantum Physics in One Dimension* (Oxford: Clarendon)
[51] Barredo D, Lienhard V, de Léséleuc S, Lahaye T and Browaeys A Synthetic three-dimensional atomic structures assembled atom by atom *Nature* **561** 79–82
[52] Wang Y, Zhang X, Corcovilos T A, Kumar A and Weiss D S 2015 Coherent addressing of individual neutral atoms in a 3D optical lattice *Phys. Rev. Lett.* **115** 043003
[53] Wang Y, Kumar A, Wu T Y and Weiss D S 2016 Single-qubit gates based on targeted phase shifts in a 3D neutral atom array *Science* **352** 1562–5
[54] Ke Y, Poshakinskiy A V, Lee C, Kivshar Y S and Poddubny A N 2019 Inelastic scattering of photon pairs in qubit arrays with subradiant states *Phys. Rev. Lett.* **123** 253601

Chapter 5

Cooperative light scattering

Light scattering from the quantum emitters provides useful information of their collective properties. Distinctive features of cooperative light scattering thus contrast the scattering patterns from uncorrelated atoms and can further shed new light on the correlated properties of light–matter interacting systems. I focus on the far-field light scattering properties of some superradiant or subradiant modes and study one-dimensional atomic array and a single atomic ring in particular.

5.1 Far-field light scattering

Here I start from a general far-field scattering formula of the light intensity with unequal times from many quantum emitters in the Heisenberg picture [1],

$$\langle \vec{E}^*(\mathbf{R}, t')\vec{E}(\mathbf{R}, t)\rangle = \left(\frac{\omega_{eg}^2}{4\pi\epsilon_0 c^2}\right)^2 \frac{|\vec{p}\,|^2}{R^2}[1 - (\hat{\mathbf{R}} \cdot \hat{\mathbf{p}})^2] \sum_{\alpha,\beta=1}^{N} e^{i\mathbf{k}_R\cdot\mathbf{r}_{\alpha\beta}}\left\langle \hat{\sigma}_\alpha^+(t')\hat{\sigma}_\beta^-(t)\right\rangle, \quad (5.1)$$

where ω_{eg} is the energy difference between the excited and ground states, the observer distance is $R = |\mathbf{R}|$, the relative position vector of the emitters is $\mathbf{r}_{\alpha\beta} = \mathbf{r}_\alpha - \mathbf{r}_\beta$, and the dipole orientation is $\vec{p}$ which is determined by the polarization of light excitations. To obtain this expression, it is assumed that the observation point is far enough from the size of the emitters, leading to the requirement of $\omega_{eg}|\mathbf{R} - \mathbf{r}_\alpha|/c \gg 1$. This results in the radiation wave vector $\mathbf{k} \approx \mathbf{k}_R \,//\, \mathbf{R}_\alpha\, [= (\mathbf{R} - \mathbf{r}_\alpha)]$ in figure 5.1 and the retarded phase $e^{ik_R(R_\alpha - R)} \approx e^{-i\mathbf{k}_R\cdot\mathbf{r}_\alpha}$. A more general expression can also be derived for a multi-level system [2] (for example of strontium atoms with three Zeeman levels in the $J = 1$ excited state), which makes equation (5.1) a tensor form of dipole transitions.

I further set the time $t = t'$ in equation (5.1) and calculate the radiation field intensity in terms of the dipole operators in the Schrödinger picture, that is, $\left\langle \Psi(t)|\hat{\sigma}_\alpha^+\hat{\sigma}_\beta^-|\Psi(t)\right\rangle$. In a Hilbert space of single excitations, I use ring structures of

doi:10.1088/978-0-7503-2699-5ch5

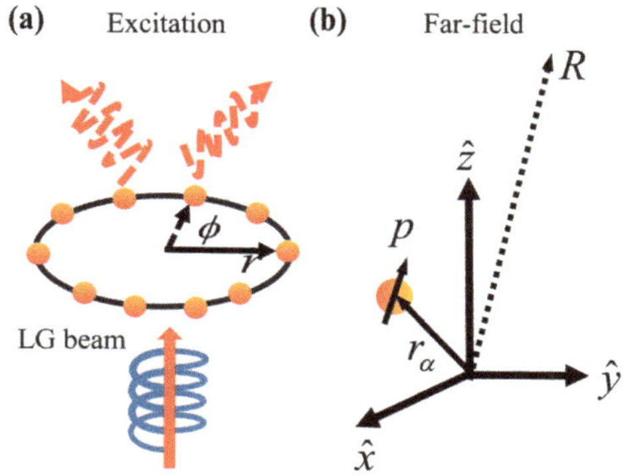

Figure 5.1. A single ring structure and far-field light scattering. (a) A single ring is demonstrated where atoms are located on the ring with equal inter-atomic arc lengths $r\phi$. Phase gradient can be imprinted via photons with finite orbital angular momenta (OAM). (b) Far-field light scattering is considered for a distant observation position $\mathbf{R}$ from the atoms locating at $\mathbf{r}_\alpha$ with a dipole moment $\mathbf{p}$.

atoms, for example, to present the phase imprinted states via photons carrying OAM (orbital angular momenta l), so-called HPI (helical phase imprinted) states $|\Psi(t)\rangle = h_l(t)|\Phi_l\rangle_{\rm HPI}$ [3], where $h_l(t)$ denotes the probability amplitude for specific $|\Phi_l\rangle_{\rm HPI}$ and

$$|\Phi_l\rangle_{\rm HPI} = \frac{1}{\sqrt{N}}\sum_{\mu_\phi=1}^{N} e^{i\mathbf{k}_L\cdot\mathbf{r}_\mu}e^{i\frac{2l\pi}{N}(\mu-1)}|\psi_\mu\rangle. \tag{5.2}$$

The above presents that N atoms sit on a single ring with a constant separation between their nearest neighbors, and the light exciting along the axis of the ring imprints the phase of $\phi = 2\pi l(\mu - 1)/N$ on the atoms, as shown in figure 5.1. This is the equivalent to the phase imprinted states of a one-dimensional array via external magnetic field gradients. A more detailed discussion on preparations of HPI states and associated paraxial approximation of Laguerre–Gaussian (LG) beams can be found in [3].

We then obtain

$$\frac{\langle \vec{E}^*(\mathbf{R}, t)\vec{E}(\mathbf{R}, t)\rangle_l}{I_0(t)} = [1 - (\hat{\mathbf{R}}\cdot\hat{\mathbf{p}})^2]\sum_{\alpha,\beta=1}^{N} e^{i(\mathbf{k}_R-\mathbf{k}_L)\cdot\mathbf{r}_{\alpha\beta}}\frac{1}{N}e^{\frac{i2l\pi}{N}(\beta-\alpha)}, \tag{5.3}$$

where $I_0(t) = I_n h_l^*(t)h_l(t)$ is the fluorescence intensity with $I_n \equiv (\omega_{eg}|\vec{p}\,|)^2/(4\pi\epsilon_0 c^2 R)^2$. The above shows a particular dependence on the geometrical distributions of atoms $\mathbf{r}_{\alpha\beta} \equiv \mathbf{r}_\alpha - \mathbf{r}_\beta$, which affects the far-field emission patterns. Therefore, the far-field emission carries the information of detailed atomic structures. Furthermore with extra phases we imprint via the lth OAM of light on the atoms, controllable far-field patterns that can be investigated to study super- and sub-radiant light scattering.

Recent studies of cooperative light scattering [4–6] have made some progress in understanding the effect of resonant dipole–dipole interactions. A subradiant light scattering can be recorded after a superradiant burst of emission [4] on the level of a thousandth or lower of an initial light intensity. The scattered light is repeatedly measured from a weakly excited cold and dense cloud. When the trap of the atoms is turned off, the light excitation (a pulse train) probes the cloud from a dense to a dilute one during expansion. This measurement thus provides important information of light scattering on the atomic densities and excitation detunings. Side scattering of light is also spotted for a subradiant emission [5], and a sub-wavelength two-dimensional atomic array can present a perfect mirror with a unit reflection [6] in a thermodynamic limit. Below, I present some rudimentary results from simple atomic structures, from which we can get some insight into cooperative light scattering.

5.2 Light scattering from two atoms

Here I use an example of two atoms sitting on a line with a distance d, which are weakly excited by a single photon with an external phase gradient along $\hat{z}$. The eigenvalues can be solved analytically from the coupling matrix in equation (2.6) in Chapter 2, which are

$$\lambda_{1,2} = -\frac{\Gamma}{2} \pm \left[\frac{\gamma_{12}(\xi) - i2\Omega_{12}(\xi)}{2}\right], \tag{5.4}$$

where Γ is the natural decay rate of the excited state, and $\xi = |\mathbf{k}_L|d$. We plot the real part of the eigenvalues in figure 5.2, which approach 2 and 0 as $d \to 0$, representing the super- and subradiant modes of Dicke radiations. For a larger d, both eigenvalues approach $\Gamma/2$, representing the independent or noninteracting regime of quantum emitters.

Next we define the normalized far-field scattering intensity,

$$\Omega_l(\theta, \phi) \equiv \langle \vec{E}^*(\mathbf{R}, t)\vec{E}(\mathbf{R}, t)\rangle_l / I_0(t), \tag{5.5}$$

and from equation (5.3), we obtain

$$\Omega_l(\theta, \phi) = (1 - \sin^2\theta\cos^2\phi)[1 + \cos(|\mathbf{k}_L|d(1 - \cos\theta) + l\pi)], \tag{5.6}$$

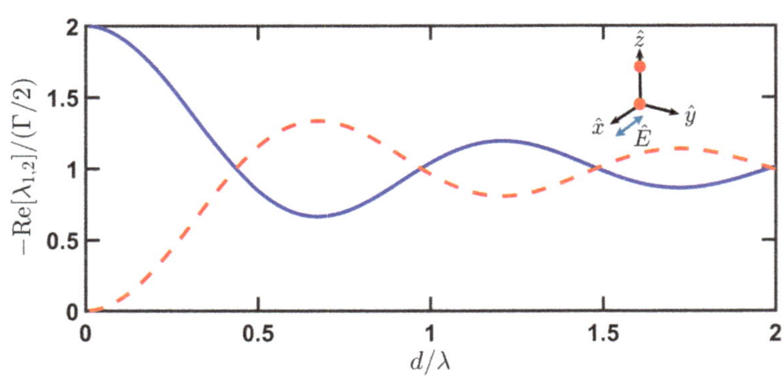

Figure 5.2. Eigenvalues for two atoms sitting on the $\hat{z}$-axis. We show the real part of the eigenvalues, $-\lambda_{1,2}/(\Gamma/2)$ for two atoms separated by d, with $\hat{x}$ polarized light excitations $\hat{E}$.

where l represents a linear phase gradient on this atomic array, resembling DM states introduced in Chapter 3. We show an example of scattering light in figure 5.3 at a small distance d. When $d \to 0$, $\Omega_l(\theta, \phi) \propto [1 + (-1)^l]$, which suggests that light is not scattered at all for an odd l. According to equation (3.8), the DM states are

$$|\Phi_l\rangle_{\rm DM} = \frac{1}{\sqrt{2}}(e^{i\mathbf{k}_L\cdot\mathbf{r}_1}|\psi_1\rangle + e^{i\mathbf{k}_L\cdot\mathbf{r}_2}e^{il\pi}|\psi_2\rangle), \tag{5.7}$$

which is a superradiant (subradiant) state for even (odd) l for $d/\lambda \ll 1$.

In figure 5.3, we show a contrasted light scattering from the super- and subradiant states, respectively. A subradiant light scattering presents a preferred backward scattering around $\theta = \pi$, whereas a superradiant scattering shows a forward scattering near $\theta = 0$. The subradiant state of $l = 1$ is generated with an imprinted π phase, which resembles an antisymmetric state (singlet state) in equation (5.7). For the dependence of azimuthal angles, we expect a maximal scattering at $\phi = \pi/2, 3\pi/2$ for both sub- and superradiant scattering. This can be clearly seen in equation (5.6), where $\cos\phi \to 0$. We note that the light scattering also depends on the atomic separation d, which modifies the scattering patterns significantly. For a larger $d \gg \lambda$, an enhancement or inhibition of decays of light scattering becomes less obvious, approaching an independent regime of atoms with negligible resonant dipole–dipole interactions. Next, we demonstrate some examples with more atoms, where a multi-atom effect on the light scattering with phase interferences will show up.

5.3 One-dimensional array

We here increase N to present its effect on the cooperative light scattering. Again we use 1D atomic array as an example, which aligns along the propagation direction. Therefore, the traveling phases along with external imprinting phases jointly determine the scattering patterns.

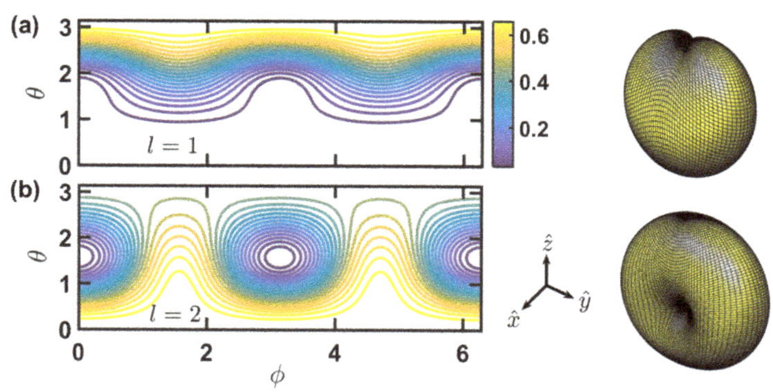

Figure 5.3. Far-field emission $\Omega_l(\theta, \phi)$ for two atoms sitting on the $\hat{z}$-axis at $d/\lambda = 0.1$. A 2D plot of $\Omega_l(\theta, \phi)$ is shown for the phase-imprinted states of (a) $l = 1$ and (b) 2 along with three-dimensional versions, which corresponds to the sub- and super-radiant light scattering respectively. $\hat{x}$ polarized light excitations are used.

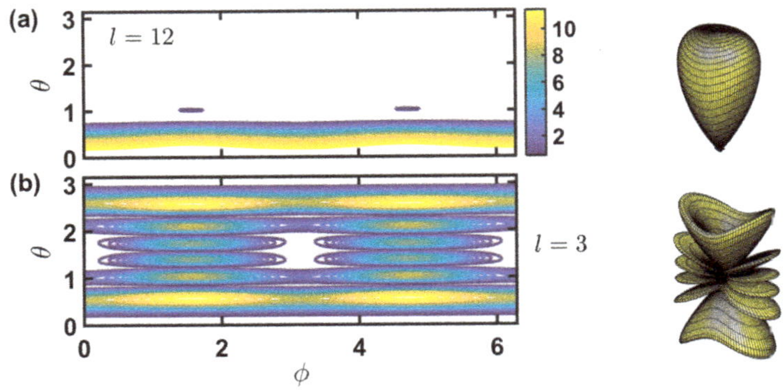

Figure 5.4. Far-field emission $\Omega_l(\theta, \phi)$ for $N = 12$ atoms sitting on the $\hat{z}$-axis at $d/\lambda = 0.25$. A 2D plot of $\Omega_l(\theta, \phi)$ is shown for the phase-imprinted states of (a) $l = 12$ and (b) 3 along with three-dimensional versions on the right, which corresponds to the super- and sub-radiant light scattering respectively. $\hat{x}$ polarized light excitations are used.

In figure 5.4, we select a super- and subradiant state with $l = 12$ and 3, respectively, to show the features of collective light scattering. The subradiant state is chosen as the most subradiant one that has the maximal overlap with the eigenstate of the lowest decay rate. As expected in multi-atom systems, a forward scattering of superradiance dominates. By contrast, a side and backward light scattering emerges in the most subradiant state as shown in figure 5.4(b). Specifically, with an equal distance of a quarter wavelength $d/\lambda = 0.25$, a traveling phase and an imprinted one with $l/N = 0.25$ exerts a multiple of $e^{i\pi/2}$ on the atomic array along $\hat{z}$, respectively, which leads to a multiple of phases $e^{i\pi}$ on the far-field scattering along $\hat{z}$. $\Omega_l(\theta, \phi)$ becomes vanishing for an even N and decreases as $N \gg 1$, which presents the most subradiant light scattering with negligible forward and backward light scatterings at $\theta = 0$ and π respectively.

Another distinctive feature in figure 5.4 shows in the peaks of side scattering around $\phi = \pi/2$ and $3\pi/2$. A total number of twelve peaks exactly reflects the total number of atoms. We will show this feature of light scattering in ring structures as well in the next section, which is a distinctive property of subradiance from superradiance and reflects the sensitivity of the imprinted phases on the far-field scattering patterns for the subradiant modes.

5.4 Single ring

For ring structures, we use the light excitation carrying OAM that imprints azimuthally increasing phases on the ring to create the collective states with correlated phases in equation (5.2). If we unfold this single ring into a line, the states with imprinted phases are exactly in the same form of an atomic array with linearly increasing phases. The difference between them lies on the geometrical distributions of atoms, leading to different traveling phases.

In figure 5.5, we use the same N and light polarizations in the light scattering from a single ring to compare with the cases of atomic array. In figure 5.5(a), the superradiant scattering is maximal in the forward and backward directions. Since we

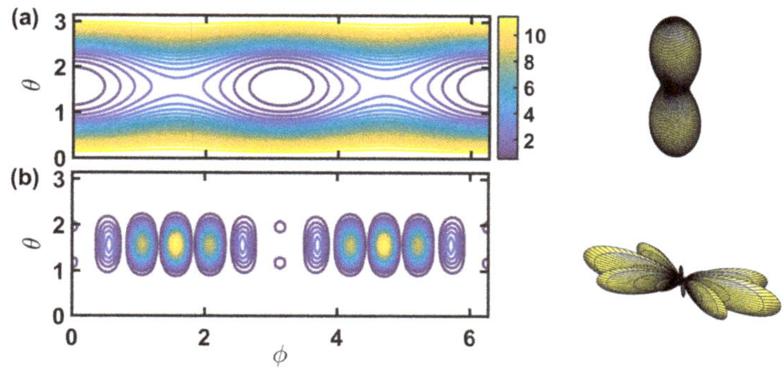

Figure 5.5. Far-field emission $\Omega_l(\theta, \phi)$ for $N = 12$ atoms sitting on a ring at $r/\lambda = 0.25$ in the $\hat{x}$–$\hat{y}$ plane. A 2D plot of $\Omega_l(\theta, \phi)$ is shown for the phase-imprinted states of (a) $l = 12$ and (b) 6 along with three-dimensional versions on the right, which corresponds to the super- and sub-radiant light scattering respectively. $\hat{x}$ polarized light excitations are used.

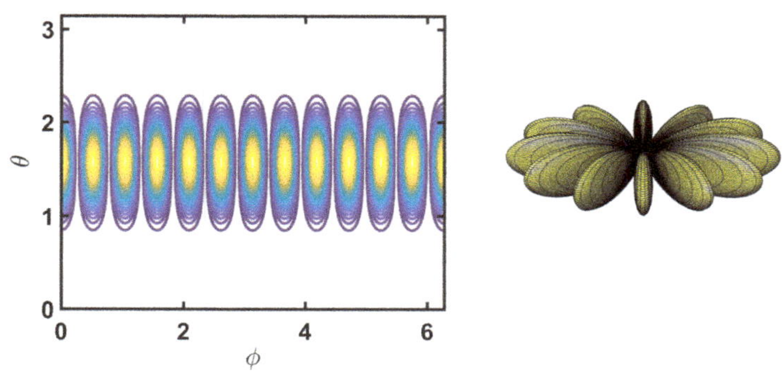

Figure 5.6. Far-field emission $\Omega_l(\theta, \phi)$ for $N = 12$ atoms sitting on a ring at $r/\lambda = 0.25$ with circular polarizations. A 2D plot of $\Omega_l(\theta, \phi)$ is shown for the phase-imprinted states of $l = 6$ along with three-dimensional versions on the right, which corresponds to the sub-radiant light scattering with symmetric scattering patterns.

put this ring on the $\hat{x}$–$\hat{y}$ plane, there is no contribution from the traveling phases. Therefore, backward scattering is also allowed in the superradiant state.

In contrast, for the most subradiant state demonstrated in figure 5.5, a purely side scattering around $\theta = \pi/2$ emerges at $l = N/2$. The scattering pattern resembles the wings of a butterfly with again the same number of peaks of a total N, though a relatively dimmed scattering intensity around $\phi = \pi$ and 2π. This vanishing intensity highly correlates to the light polarizations of $\hat{x}$, which originates from a dependence of $\cos\phi$ in the case of two atoms.

Lastly, I show an example of using light with circular polarizations in figure 5.6. Circular polarizations can be expressed in terms of $(\hat{x} \pm \hat{y})/\sqrt{2}$ in the $\hat{x}$–$\hat{y}$ plane. The symmetric scattering patterns can be clearly seen and it is evident twelve scattering peaks are evenly distributed around $\theta = \pi$. This most subradiant state shows a side scattering pattern with distinctive peaks, which depend on N, and thus further provides an extra information of number of quantum emitters due to the phase interferences.

More complex ring structures, for example, stacked rings along $\hat{z}$ or multiple concentric rings [3], can show different scattering patterns influenced by traveling phases or imprinted ones, respectively. Subradiant light scattering with a characteristic $\Gamma_f < \Gamma/2$ can change to a superradiant one $\Gamma_f > \Gamma/2$ when more rings are involved for the same l, showing the multi-atom effect. Other light polarizations (radial and azimuthal) can also be used to excite these ring structures, showing characteristics that can be distinguished from each other [3]. Cooperative multi-photon scattering can also be investigated to show a directional subradiance [7].

References

[1] Lehmberg R H 1970 Radiation from an N-atom system. I. General formalism *Phys. Rev.* A **2** 883–8

[2] James D F V 1993 Frequency shifts in spontaneous emission from two interacting atoms *Phys. Rev.* A **47** 1336–46

[3] Jen H H, Chang M S and Chen Y C 2018 Cooperative light scattering from helical-phase-imprinted atomic rings *Sci. Rep.* **8** 9570

[4] Guerin W, Araújo M O and Kaiser R 2016 Subradiance in a large cloud of cold atoms *Phys. Rev. Lett.* **116** 083601

[5] Bromley S L *et al* 2016 Collective atomic scattering and motional effects in a dense coherent medium *Nat. Commun.* **7** 11039

[6] Shahmoon E, Wild D S, Lukin M D and Yelin S F 2017 Cooperative resonances in light scattering from two-dimensional atomic arrays *Phys. Rev. Lett.* **118** 113601

[7] Jen H H 2018 Directional subradiance from helical-phase-imprinted multiphoton states *Sci. Rep.* **8** 7163

IOP Publishing

Collective Light Emission
Many quantum emitters
Hsiang-Hua Jen

Chapter 6

Subradiance in a chirally coupled atomic chain

An atomic chain with chiral couplings can be realized in atom–nanofiber or atom–waveguide systems, forming one of the strongly coupled and correlated light–matter interfaces. Subradiance, in particular, emerging from such a platform can lead to many interesting and distinctive features that conventional quantum optics cannot achieve. In this chapter, I present some recent research topics in a one-dimensional system of quantum emitters with nonreciprocal couplings, which are particularly relevant to the collective radiations in nanophotonics platforms.

6.1 A chiral coupling matrix

Many chiral objects actually exist in our Mother Nature. In addition to mathematical objects or chiral molecules (amino acid, for example), our two hands show one example of chiral objects. They are mirror images to each other but are not able to be superposed. Taking massless particles of light as another example, its two circular polarizations present the 'handedness' of light, and chirality shows in the inner product between light propagation direction ($\hat{k}$) and its handedness ($\hat{s}$). Under this condition, the helicity of light polarizations exactly reflects their chirality, depending on the signs of $\hat{k} \cdot \hat{s}$.

In conventional quantum optical systems, the spontaneous emissions from the excited quantum emitters in free space originate from reciprocal couplings, which exchange the radiations back and forth between the atoms and preserve the time-reversal symmetry. This also reflects in the resonant dipole–dipole interaction form of equations (2.7) and (2.8), for example, where the interaction strengths mainly depend on the interparticle distance instead of coupling directions. By contrast, a chirally coupled interface can be made in the evanescent waves [1, 2] at the total internal reflection from the guided photons transmitted in fibers. An exponential reduction of evanescent waves in the direction perpendicular to the propagation leads to a finite transverse spin angular momentum of light and the longitudinal component of light polarizations. And as such, the spin and the linear momentum of

doi:10.1088/978-0-7503-2699-5ch6

light are locked with each other, and this spin-momentum locking plays an essential element to realize the chirally coupled systems.

We take the minimal and effective model of a chirally coupled atomic chain [3], which considers nonreciprocal decay channels between left- and right-propagating emissions as shown in figure 6.1. In Lindblad forms for arbitrary observables Q, we have

$$\frac{dQ}{dt} = -i[Q, H_L + H_R] + \mathcal{L}_L[Q] + \mathcal{L}_R[Q], \tag{6.1}$$

where

$$H_L \equiv -\frac{i\hbar\gamma_L}{2}\sum_{\mu<\nu}\left(e^{ik|x_\mu - x_\nu|}\sigma_\mu^\dagger\sigma_\nu - \text{H.c.}\right), \tag{6.2}$$

$$H_R \equiv -\frac{i\hbar\gamma_R}{2}\sum_{\mu>\nu}\left(e^{ik|x_\mu - x_\nu|}\sigma_\mu^\dagger\sigma_\nu - \text{H.c.}\right), \tag{6.3}$$

describe the energy shift of 1D RDDI, and the Lindblad forms

$$\mathcal{L}_L[\hat{Q}] \equiv -\frac{\gamma_L}{2}\sum_{\mu,\nu}\left\{e^{-ik(x_\mu - x_\nu)}\left(\sigma_\mu^\dagger\sigma_\nu Q + Q\sigma_\mu^\dagger\sigma_\nu - 2\sigma_\mu^\dagger Q\sigma_\nu\right)\right\}, \tag{6.4}$$

$$\mathcal{L}_R[\hat{Q}] \equiv -\frac{\gamma_R}{2}\sum_{\mu,\nu}\left\{e^{ik(x_\mu - x_\nu)}\left(\sigma_\mu^\dagger\sigma_\nu Q + Q\sigma_\mu^\dagger\sigma_\nu - 2\sigma_\mu^\dagger Q\sigma_\nu\right)\right\}, \tag{6.5}$$

characterize the collectively dissipative process.

This effective model highlights a direction-dependent frequency shift and two-channel decay rate to the left and right of the chain. Especially, this particular model includes an infinite-range dipole–dipole interaction that mediates over every

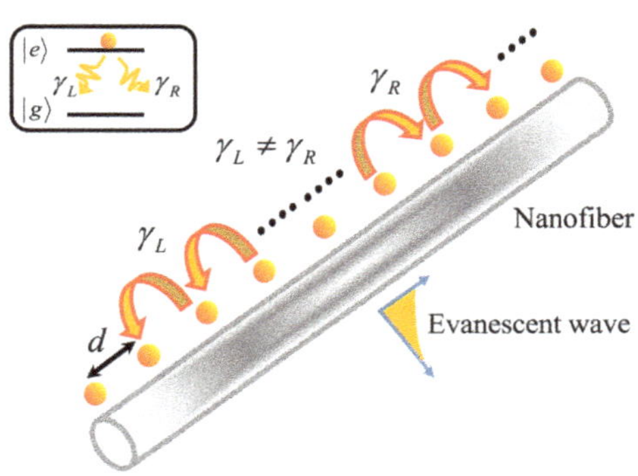

Figure 6.1. A schematic plot for 1D nanofiber coupled to an atomic array. This atom–nanofiber system offers a platform of guided emissions with chiral decay channels $\gamma_L \neq \gamma_R$. The interparticle distance is denoted as d, which characterizes the strength of 1D dipole–dipole interactions. This light–matter interface with nonreciprocal couplings is realized via evanescent waves from the surface of an optical fibers interacting with two-level quantum emitters near the fiber.

quantum emitter. This is possible only for a light–matter interface with 1D reservoirs, which I have discussed in section 2.4. Moreover, this effective Hamiltonian models the nonreciprocal couplings and even allows a description of unidirectional channel, where the emissions can be guided only in one way of the fiber.

Next I assume a Hilbert space of singly excited states. This is true when the weakly excited regime is considered, and we are able to obtain the interaction Hamiltonian V in the Schrödinger equations in terms of the bare-state basis $|\psi_\mu\rangle = |e\rangle_\mu |g\rangle^{\otimes(N-1)}$ [4],

$$V = \begin{bmatrix} -F_{11} & -F_{12}+iG_{12} & -F_{13}+iG_{13} & \dots & -F_{1N}+iG_{1N} \\ -F_{12}^*+iG_{12}^* & -F_{22} & -F_{23}+iG_{23} & \dots & -F_{2N}+iG_{2N} \\ -F_{13}^*+iG_{13}^* & -F_{23}^*+iG_{23}^* & -F_{33} & \dots & -F_{3N}+iG_{3N} \\ \vdots & \vdots & \vdots & \ddots & \vdots \\ -F_{1N}^*+iG_{1N}^* & -F_{2N}^*+iG_{2N}^* & -F_{3N}^*+iG_{3N}^* & \dots & -F_{NN} \end{bmatrix}. \quad (6.6)$$

We consider a general state function,

$$|\Psi(t)\rangle = \sum_\mu c_\mu(t)|\psi_\mu\rangle, \quad (6.7)$$

where the atomic dynamics can be determined via

$$\dot{c}_\mu = V_{\mu,\nu} c_\nu. \quad (6.8)$$

In general, $F_{\mu\nu}$ and $G_{\mu\nu}$ are complex, and $V_{\mu,\nu} = V_{\nu,\mu}$ only when $\gamma_L = \gamma_R$. Furthermore, V can be explicitly written as

$$V = \begin{bmatrix} -\frac{\gamma_L+\gamma_R}{2} & -\gamma_R e^{-ik|x_{1,2}|} & -\gamma_R e^{-ik|x_{1,3}|} & \dots & -\gamma_R e^{-ik|x_{1,N}|} \\ -\gamma_L e^{-ik|x_{1,2}|} & -\frac{\gamma_L+\gamma_R}{2} & -\gamma_R e^{-ik|x_{2,3}|} & \dots & -\gamma_R e^{-ik|x_{2,N}|} \\ -\gamma_L e^{-ik|x_{1,3}|} & -\gamma_L e^{-ik|x_{2,3}|} & -\frac{\gamma_L+\gamma_R}{2} & \dots & -\gamma_R e^{-ik|x_{3,N}|} \\ \vdots & \vdots & \vdots & \ddots & \vdots \\ -\gamma_L e^{-ik|x_{1,N}|} & -\gamma_L e^{-ik|x_{2,N}|} & -\gamma_L e^{-ik|x_{3,N}|} & \dots & -\frac{\gamma_L+\gamma_R}{2} \end{bmatrix}. \quad (6.9)$$

The above form presents a nonsymmetric form of the chiral coupling matrix, which is in huge contrast to the conventional light–matter interactions with reciprocal couplings. V is not a normal or a Hermitian matrix in general, and in particular V is a defective one, that is, V cannot be decomposed by linearly independent eigenstates. Therefore, the method of eigen-decomposition is not adequate to analyze the system dynamics, and instead we numerically solve it directly from

$$\frac{d}{dt}\vec{c} = V\vec{c}, \tag{6.10}$$

where $\vec{c} \equiv (c_1(t), c_2(t), \ldots, c_N(t))$ with given initial conditions of $\vec{c}(t = 0)$.

We note that the above formalism can be applied when some particular initialized states are known. This can be done by side excitations [5], and the system dynamics can then be evolved via waveguide-mediated decay channels. On the other hand, an input–output formalism [6–8] can also be used to investigate the photon correlations in such atom–waveguide interfaces. This, by contrast, leads to a perspective of the field propagation in optical fibers, which can go through multiple reflections and transmissions mediated by atom–waveguide couplings. Both regimes of excitations are interesting and allow the studies of distinctive quantum correlations in either atomic or photonic degrees of freedom.

6.2 Subradiance dynamics

Next I show some examples of subradiance dynamics in a chirally coupled atomic chain determined by equation (6.6) [4] with an initial uniform distribution $c_m(t = 0) = 1/\sqrt{N}$. For $N = 2$, we show the atomic populations $P_{1,2}(t) \equiv |c_{1,2}(t)|^2$ in figure 6.2 at two specific $\xi = \pi/4$ and $3\pi/4$, which respectively show early superradiant and subradiant behaviors as in the cascaded scheme ($\gamma_L = 0$). At $\xi = \pi/4$ in figure 6.2(a), both $P_{1,2}(t)$ decay more superradiantly as γ_L increases, and $P_{tot}(t)$ in the inset approaches $e^{-(2+\sqrt{2})\gamma t}$ when $\gamma_L \to \gamma_R$, significantly faster than $e^{-2\gamma t}$ in the noninteracting regime. This enhancement of superradiance is expected for uniform excitations and finite γ_L at small ξ or $\xi \sim 2\pi$, where all atoms are in phase as in Dicke's superradiant regime. We note of the same trend of shortened lifetime as γ_L increases for $\xi = 0$ and $\pi/2$ as in figure 6.2(a).

Interestingly, a repopulation occurs from a signature of a bump in $P_2(t)$ in figures 6.2(a) and 6.2(b), especially for $\gamma_L \sim 0$. This presents a sequential

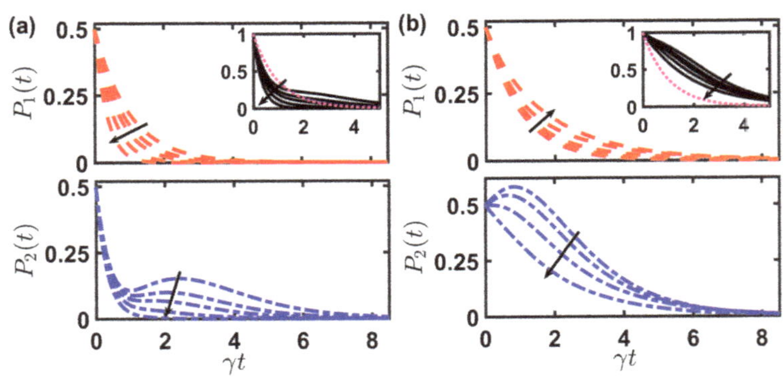

Figure 6.2. Excited state populations of $N = 2$ for (a) $\xi = \pi/4$ and (b) $3\pi/4$ in the cascaded and non-cascaded scheme. With a constant $\gamma_R = \gamma$, the arrows indicate the nonreciprocal couplings from $\gamma_L/\gamma = 0$, 0.1, 0.2, 0.4, to 0.8. The first two lines of $P_2(t)$ in (b) overlap each other. The line of $e^{-\gamma t}$ (magenta dotted line) in both insets is plotted for comparison.

transmission of light under unidirectional couplings, in contrast to the regime with reciprocal couplings which transfer light to both sides on an equal footing.

Similar to the enhanced superradiance in figure 6.2(a), we have a less subradiant $P_{tot}(t)$ at $\xi = 3\pi/4$. This can be seen in figure 6.2(b) and the inset where, as γ_L increases, the second atom radiates faster and overcomes the increasing lifetime of the first one. As $\gamma_L \to \gamma_R$, $P_{tot}(t)$ approaches $e^{-(2-\sqrt{2})\gamma t}$, still subradiant compared to $e^{-\gamma t}$. For the case of reciprocal couplings when $\gamma_R = \gamma_L$, we are able to derive, in general for arbitrary ξ, the eigenvalues $\beta/\gamma = -e^{-i\xi}(1 + e^{i\xi})$ for the eigenstate of the uniform excitations. This eigenvalue becomes the most superradiant at $\xi = 0$, that is $\beta = -2\gamma$, and again it becomes the most subradiant as $\beta = 0$ at $\xi = \pi$.

In figure 6.3, we further show the radiation dynamics in the non-cascaded scheme at $\xi = \pi$ under the most subradiant coupling regime. This most subradiant coupling regime presents an alternating positive and negative phases in the couplings between the constituent atoms, leading to an effective light trapping. This π phase change also resembles a singlet state of $N = 2$, an antisymmetric state configuration of Dicke's subradiance. Here I first present few atom cases. For both cases of $N = 2$ and 3, all atomic excitations prolong in time as γ_L increases, making the system more and more subradiant.

For $N = 3$ at the early stage in figure 6.3(b), the atoms on the edges (the first and the third) decay faster since they couple with each other with losses, while they can be repumped by the radiation from the central one. As time evolves, the central atom decays and transfers the excitation more to the right than the left since $\gamma_R > \gamma_L$. For an even longer time, these atomic populations start to oscillate back and forth due to interferences of light transmissions and reflections. These light exchanges or scatterings play important roles in determining the radiation evolutions in the 1D chirally coupled system and manifest different decay behaviors when N is even or odd, which I will discuss below for a longer atomic chain. The alternating couplings in the most subradiant coupling regime can be clearly seen from the early evolutions of populations at $t \gtrsim 0$ in figure 6.3(b), where $P_1(t)$ and $P_3(t)$ share the same trend of population decrease.

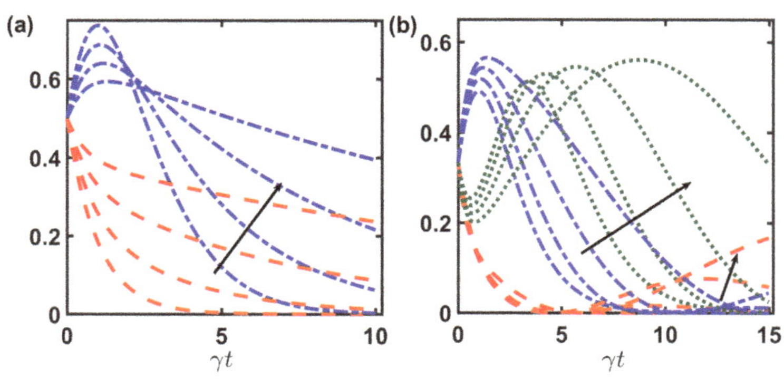

Figure 6.3. Excited state populations of (a) $N = 2$ and (b) $N = 3$ for $\xi = \pi$ in the cascaded and non-cascaded scheme. With a constant $\gamma_R = \gamma$, the arrows indicate the nonreciprocal couplings from $\gamma_L/\gamma = 0$, 0.2, 0.4, to 0.6. The respective line symbols are the same as figure 6.2 with $P_3(t)$ (green dotted line).

For a longer chirally coupled atomic chain, we expect multiatom interferences on the decay behaviors. The directionality factor of $D \equiv (\gamma_R - \gamma_L)/(\gamma_R + \gamma_L)$ [5] is defined, which is between 0 and 1 when γ_L goes from $\gamma_L = \gamma_R$ to 0. A vanishing D indicates the reciprocal couplings where the radiations are symmetric in both directions, while a $D = \pm 1$ represents a unidirectional radiation. The results for the region of $D = [-1,\ \ 0]$ can be obtained by exchanging $\gamma_R \leftrightarrow \gamma_L$. Here I focus again on the most subradiant coupling regime at $\xi = \pi$. In figure 6.4, we characterize the subradiant decay constants Γ_f by fitting the time evolutions of a total number of excited-state population,

$$P_{tot}(t) \equiv \sum_m^N |c_m(t)|^2, \tag{6.11}$$

to an exponential function, which provides a timescale for comparisons. As $D \to 0$, the system becomes decoherence-free, where finite steady-state excitations are sustained. As N increases, we see the reduction of Γ_f, a signature of enhanced quantum coherence in many-atom systems, which is also present in 1D open atomic arrays [9]. The scaling of N for respective D can be extrapolated from figure 6.4 in a form of $N^{-\alpha}$, where α increases as D increases. We find that a unidirectional coupling of $D = 1$ allows a stronger dependence on the number of atoms, which leads to a faster drop of Γ_f as N increases. It would dominate the non-cascaded coupling, for example of $D \geqslant 0.5$, when $N \gtrsim 10^2$, where $\Gamma_f \sim 3 \times 10^{-3}\gamma$ can be reached.

As N increases, we find a relatively flat dependence of Γ_f between $N = 2m$ and $2m + 1$, making them a separate group of decay behaviors for different integers m. We highlight the clear regions of these groups by a circle in figure 6.4, which indicates a significant change of decay behaviors only within $N = 2m$ or $2m + 1$ with integers m. This separation is more obvious when $D \lesssim 0.5$, where the relative deviation, $|\Gamma_f^{2m} - \Gamma_f^{2m+1}|/(\Gamma_f^{2m} + \Gamma_f^{2m+1})$, decreases as N increases. In contrast, for a larger D, it is smoothed out. From this feature of separate groups of fitted

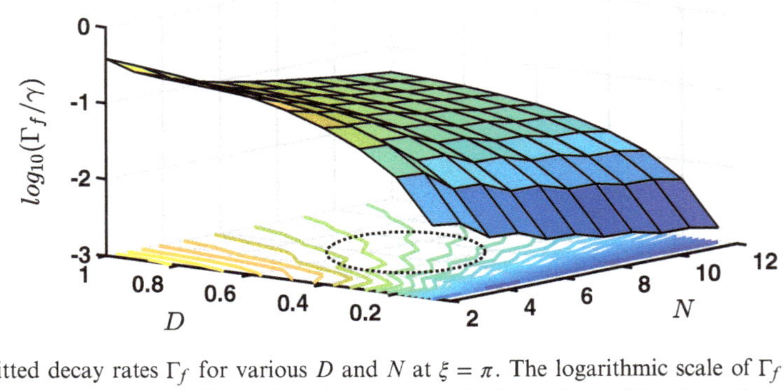

Figure 6.4. Fitted decay rates Γ_f for various D and N at $\xi = \pi$. The logarithmic scale of Γ_f decreases as D approaches 0 and as N increases. The dashed circle denotes the flat regions with a dependence of N, showing similar decay behaviors for $N = 2m$ and $2m + 1$ with an integer m.

macroscopic lifetimes, it shows another qualitatively different decay behavior in the chirally coupled system for different left and right decay channels.

As a final remark in this section, there is a plethora of unique subradiance dynamics allowed under nonreciprocal couplings. In such chirally coupled atomic chain [4], a sequential radiation can lead to excitation plateaus, a displaced atom can lead to reemergence of excitation plateaus, and a persistent localized excitation can also be possible. This offers many possibilities to manipulate atomic states and create dimer-like or trimerized states. Exotic many-body spin dynamics can be potentially simulated via infinite dipole–dipole interactions near atom–nanofiber or atom–waveguide systems.

The subradiance dynamics presented in this section, however, can suffer from nonguided couplings or nonradiative losses. Therefore, a significantly strong coupling between light and atoms should be required to reveal clear features of long time subradiance. An external cavity integrated in waveguides or optical fibers can help enhance the coupling, and this effort has continued to advance to reach strong coupling regimes. The advantage of neutral atoms near nanofibers is their identical properties as quantum emitters, whereas for quantum dots in waveguides, they can suffer from inhomogeneous fabrication processes. Nevertheless, quantum dots can be externally tuned to behave as if they are identical to keep quantum coherences intact. Each platform faces various issues and challenges which may not be on the same level, but respective advantages give hope to being potentially the best quantum interface composed of only a few quantum emitters. This also can be scaled up to form a quantum network or a prototype of quantum computation.

6.3 Quantum-coherence-enhanced subradiance

Here I release the initial condition of uniform excitations, and focus on the end excitation in the left of the chain with a number of N_i [10] under the most subradiant coupling regime ($\xi = \pi$). The initialized state can be written as

$$|\Psi(t=0)\rangle = \sum_{m=1}^{N_i} \frac{1}{\sqrt{N_i}} \hat{\sigma}_m^\dagger |0\rangle^{\otimes N}. \tag{6.12}$$

The above is a well-known W state that is maximally entangled within a dimension of $N_i > 1$[1]. In figure 6.5, we show the examples of cascaded ($\gamma_L = 0$) and noncascaded ($\gamma_L \neq 0$) coupling regimes, where a subradiant time evolution can be made slower as N_i increases. We therefore call it quantum-coherence-enhanced subradiance since the coherence or entanglement of the initialized state is greater when more N_i atoms are involved.

For the noncascaded regime in figure 6.5(b), a more subradiant evolution is allowed, showing a signature of interference between left- and right-propagating fields. One common feature in a chirally coupled atomic chain is the excitation

[1] The entropy of entanglement of W state can be calculated as $\ln N_i$ from a density matrix $|\Psi\rangle\langle\Psi|$, which increases monotonically as N_i increases.

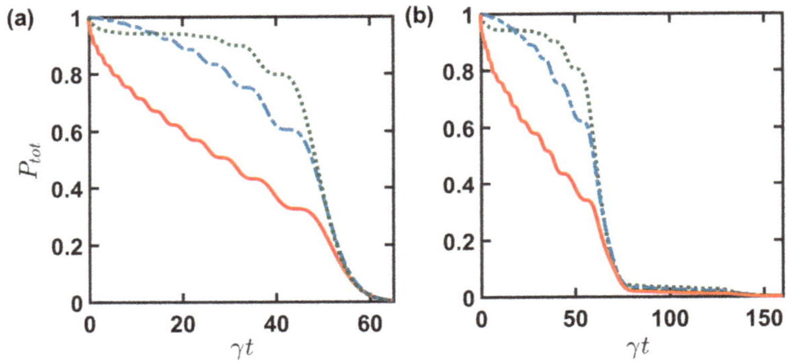

Figure 6.5. Total excited state populations in the cascaded and noncascaded schemes for (a) $\gamma_L = 0$ and (b) 0.2γ, at $\xi = \pi$ with $\gamma_R = \gamma$. Time evolutions of the total excited state populations are plotted for $N = 15$ with initial end excitations $N_i = 1$ (red solid line), 2 (blue dashed line), and 3 (green dotted line).

plateau in figure 6.5, making the decay non-exponential in general. When γ_L becomes finite in the noncascaded scheme as in figure 6.5(b), an even larger plateau shows up at $\gamma t \sim 100$. This flat region of excitation will result in a burst of light radiation [4] as time evolves, which can be an indicator of ordering excitations in the atomic chain.

Many other ways of excitations can be studied, for example, of central excitations [10]. Enhanced subradiance can also be achieved as N_i increases. The clear excitation plateaus can be revealed only when position fluctuations of atoms are made small. Therefore, the assumption of the equidistant atomic array is stringent, but is not entirely impossible in tight lattice potentials or confined microtraps. As the fluctuations increase, the cascaded scheme sustains the form without fluctuations compared to the noncascaded scheme. This is because the couplings of both sides are more seriously affected by position fluctuations, whereas the cascaded scheme with unidirectional coupling withstands them better [10].

6.4 Steady states under weak excitations

In the last section, I introduced some results of steady states in chirally coupled quantum emitters under weak excitations. Under this driven-dissipative regime, the interaction of equation (6.6) is modified to include a diagonal term of $-i\Omega$, where Ω is the Rabi frequency applied to the emitters. This term originates from a coupling of $i\Omega(\hat{\sigma}_{ee} - \hat{\sigma}_{gg})$ with a population difference between the excited and the ground states. Since a weak excitation regime is considered, it can then be simplified to a constant pump to the atomic coherence $\hat{\sigma}_\mu$ or the state function of $|\psi_\mu\rangle$ in equation (6.7).

The complete steady-state phase diagram of a chirally coupled atomic chain has been obtained [11]. This work presents a detailed study of various interaction-driven steady states with competing parameters between directionality D and interatomic distance ξ. D quantifies a flow of coupling between the left and right decay channels, while ξ characterizes the dipole–dipole interactions that can lead either to super-radiant or subradiant coupling regimes. These two competing interactions thus allow steady states with crystalline orders, resembling repulsive particles confined under an attractive potential, and edge or hole excitations, which, respectively, show the localized or extended features of the atomic excitations.

Under weak excitations, we use normalized excitation populations to present the steady state distributions,

$$\tilde{P}_j = \frac{|c_j^{(s)}|^2}{\sum_{j=1}^{N} |c_j^{(s)}|^2}, \tag{6.13}$$

where s stands for steady-state. The crystalline orders can be identified by finite structure factors,

$$S(k) = \sum_{j,m}^{N} e^{ik(j-m)} \tilde{P}_j \tilde{P}_m, \tag{6.14}$$

and the localized or extended features of the states can be characterized by a participation ratio (PR),

$$\mathrm{PR} \equiv \frac{\left(\sum_{j=1}^{N} \Delta\tilde{P}_j\right)^2}{\sum_{j=1}^{N} \left(\Delta\tilde{P}_j\right)^2}. \tag{6.15}$$

A referenced state of uniform distributions N^{-1} at every sites is used in $\Delta\tilde{P}_j = |\tilde{P}_j - N^{-1}|\Theta(\tilde{P}_j - N^{-1})$, where the Heaviside step function Θ evaluates the difference from an extended state with uniform distributions. When PR is low, the states are quite localized, in contrast to a large PR, which indicates a nonlocalized state.

In figure 6.6, the states with crystalline orders are shown. We can see the clear features of periodic occurrences of atomic excitations, which leads to finite structure factors. In these parameter regimes, a unidirectional coupling drives the system to populate preferentially to the right, but a counter force from 1D dipole–dipole interactions seems to balance this unidirectional coupling to form a crystalline order.

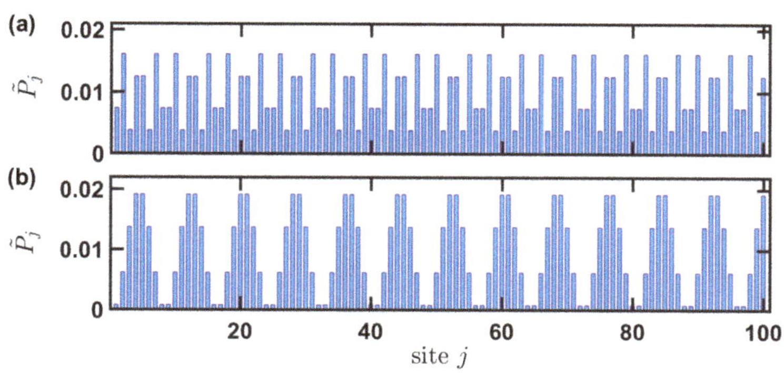

Figure 6.6. Steady states with crystalline orders. The parameters are chosen at $D = 1$ and (a) $\xi = \pi/4$, (b) $\xi = 3\pi/4$, respectively, with $N = 100$.

For the examples of edge or hole excitations, we present in figure 6.7 where two edge or hole excitations are populated. These steady-state phases are identified when $\tilde{P}_{1,N} > N^{-1}$ or $\tilde{P}_{1,N} < N^{-1}$, respectively. The properties of PR can be straightforwardly calculated as 2 and 98, which respectively reveals two equally and locally populated excitations and two hole ones uniformly melted in the bulk. At $D = 0$, these two phases are thermodynamically stable since the PR of the former is always two, and for the latter, PR is proportional to $(N - 2)$, an extended feature for hole excitations.

In contrast to the above stable phases, a region of two kinds of steady states is shown in figure 6.8. At such subradiant coupling regime of $\xi = \pi$, an extra emitter can actually change the whole distributions of the steady states. For an even N, a directionality $D > 0$ indicates a favorite coupling to the right of the chain. Therefore, figure 6.8(a) shows more populations on the right over the entire atomic chain. By contrast in figure 6.8(b), a minimum shows up near the center of the bulk, instead of the left in figure 6.8(a). This can be understood as a balanced coupling strength for an even N, where an equal number of ±1 from the phases $e^{im\xi}$ at mth

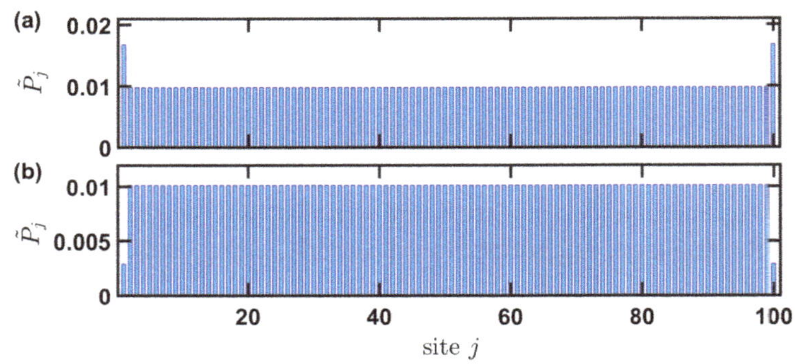

Figure 6.7. Steady states with two edge or hole excitations. The parameters are chosen at $D = 0$ and (a) $\xi = \pi/4$, (b) $\xi = 3\pi/4$, respectively, with $N = 100$.

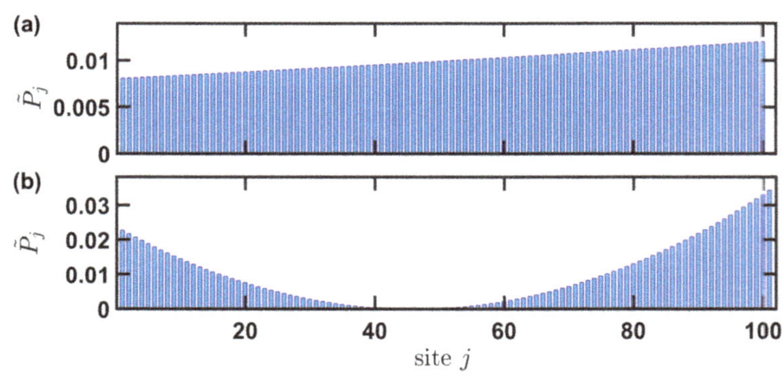

Figure 6.8. A region of steady states with N dependence. The parameters are chosen at $D = 0.1$ and $\xi = \pi$ with (a) $N = 100$ and (b) $N = 101$, respectively.

sites are present, whereas for an odd N, an extra emitter breaks the balanced coupling terms, leading to population accumulations at the edges with a minimum near the center of the bulk.

We note of two critical points allowed in such a chirally coupled atomic chain, where the time to reach steady states becomes infinite at $D = 0$ and $\xi = 0$ or $\xi = \pi$ [11]. These parameter regimes allow decoherence-free eigenmodes, and as such, long time oscillations between these modes make the system hard to reach an equilibrium. These subradiant modes play crucial roles in quantum state engineering and exotic subradiance dynamics. This may further open up other interesting research directions in nonequilibrium many-body spin dynamics [12], the creation of dimer-like states [4], or selective transport of atomic excitations [13] via an interplay between the chirality and collective light-induced dipole–dipole interactions of the system.

References

[1] Bliokh K, Bekshaev A and Nori F 2014 Extraordinary momentum and spin in evanescent waves *Nat. Commun.* **5** 3300

[2] Bliokh K Y and Nori F 2015 Transverse and longitudinal angular momenta of light *Phys. Rep.* **592** 1–38

[3] Pichler H, Ramos T, Daley A J and Zoller P 2015 Quantum optics of chiral spin networks *Phys. Rev.* A **91** 042116

[4] Jen H H, Chang M S, Lin G D and Chen Y C 2020 Subradiance dynamics in a singly excited chirally coupled atomic chain *Phys. Rev.* A **101** 023830

[5] Mitsch R, Sayrin C, Albrecht B, Schneeweiss P and Rauschenbeutel A 2014 Quantum state-controlled directional spontaneous emission of photons into a nanophotonic waveguide *Nat. Commun.* **5** 5713

[6] Shen J T and Fan S 2005 Coherent photon transport from spontaneous emission in one-dimensional waveguides *Opt. Lett.* **30** 2001–3

[7] Mahmoodian S, Čepulkovskis M, Das S, Lodahl P, Hammerer K and Sørensen A S 2018 Strongly correlated photon transport in waveguide quantum electrodynamics with weakly coupled emitters *Phys. Rev. Lett.* **121** 143601

[8] Kiilerich A H and Mølmer K 2019 Input-output theory with quantum pulses *Phys. Rev. Lett.* **123** 123604

[9] Jen H H, Chang M S and Chen Y C 2016 Cooperative single-photon subradiant states *Phys. Rev.* A **94** 013803

[10] Jen H H 2019 *Quantum-coherence-enhanced subradiance in a chiral-coupled atomic chain* (arXiv: 1-903.053-52)

[11] Jen H H 2020 Steady-state phase diagram of a weakly driven chiral-coupled atomic chain *Phys. Rev. Res* **2** 013097

[12] Hung C L, González-Tudela A, Cirac J I and Kimble H J 2016 Quantum spin dynamics with pairwise-tunable, long-range interactions *Proc. Natl. Acad. Sci.* **113** E4946–55

[13] Jen H H 2019 Selective transport of atomic excitations in a driven chiral-coupled atomic chain *J. Phys. B: At. Mol. Opt. Phys.* **52** 065502

Chapter 7

Applications in quantum information processing and quantum simulation

We discuss two important applications that are relevant to collective light emissions from many quantum emitters. The first is quantum information processing and the second is quantum simulations. These two subjects have attracted many cross-field scientists to invent new technologies and create new methodologies in versatile quantum systems. The prospects of these two applications, respectively, are to generate an operational quantum computer and to explore new physics via quantum systems with controllable Hamiltonians.

7.1 Quantum information processing

Quantum information is the foundation for many applications, and one of them is the well-known quantum computation. It contains an immense realm of research subjects, as broad as wherever quantum mechanics concerns. Here I will not start from the very beginning of how quantum information arises as a hot topic and evolves as a pervading concept in almost every quantum system [1, 2]. Instead, I point out the Nobel physics prize in 2012 as a motivation here, which was awarded to Serge Haroche and David J Wineland jointly for ground-breaking techniques in measuring and controlling individual quantum systems, which are atoms and light quanta. This presents years of effort to successfully manipulate the quantum states of photons and atoms, and an important step towards building an efficient quantum interface [3] that makes the processing of quantum information possible. One recent advanced demonstration is to generate quantum entanglement between atomic internal states and molecular rotational states [4]. This promises a flexibility of entanglement prepared in wide frequency ranges and enables a hybrid quantum system.

Since the information encoded quantum mechanically is fragile to its environment owing to dissipation or decoherence, there is a limited time of its usefulness before it decoheres. This period of time could be a neutral atom's intrinsic lifetime of its

doi:10.1088/978-0-7503-2699-5ch7

excited state or a lifetime of a quantum degenerate gas if the information is encoded in terms of spin or momentum configurations respectively. They remain quantum for a limit of time owing to quantum or ambient thermal fluctuations. The quantum system that encodes quantum information can be denoted as a qubit, and to make these qubits operational, this information must pass on or should be stored for later readout under control. Flying photons are perfect to transmit through air or fibers without much attenuation when their optimal frequencies are applied, whereas stationary atoms are best for information storage and the controllable retrieval of it.

There are many applications related to quantum technologies [5] that are waiting to mature to have huge impact on the human world. Here I focus on two applications of quantum memory and quantum state engineering, where the collective effect of light–matter interactions is most relevant. Quantum memory serves as an essential element for quantum repeater protocol and is indispensable in almost every operational platforms to be functional. Meanwhile, quantum state engineering is crucial in quantum control and quantum simulations, which is the foundation for quantum technology. Other important quantum technologies include quantum communication and quantum computation, where the former transmits information to longer distances and fulfills the architecture on a large scale, while the latter provides a next-generation computation technology that will steadily achieve its short-, mid- and long-term goals [6].

7.1.1 Quantum memory

Quantum memory has advanced its capability in versatile quantum systems, where both efficiency and lifetime are pushed to their respective optimum [7]. In a platform of cold atoms, near unit efficiency can be approached, and an order of one minute memory can be achieved [8, 9]. However, neither of both can be fulfilled to their optimum at the same time owing to their respective technical challenges, and the respective optimum compromises the performance of the other unfortunately.

There are many techniques to store quantum information. In figure 7.1, we first motivate the use of quantum memory in a quantum repeater protocol and select two platforms of electromagnetically induced transparency (EIT) and collective subradiant states as storage units. A quantum repeater is essential in distributing quantum entanglement to long distances before it decoheres, which makes it possible for other important elements in the quantum network, for example, quantum teleportation and quantum communication. It involves multiple entangled pairs between the end parties (A and B) in figure 7.1(a) and makes them entangled by projections (Bell measurement). This protocol cannot be done without efficient quantum memories with enough lifetime.

In figures 7.1(b) and 7.1(c), I respectively show two examples to store light quanta in EIT setup and two-level atoms. The former uses three-level systems where a weak probe light is transferred into a collective spin excitation to $|s\rangle$ under a strong control field operating on two empty levels, and the dark-state polariton between light fields and atomic spins can be formed [10]. This light quanta can be slowed and effectively stored in the atoms and can be retrieved later by turning the control field back. This

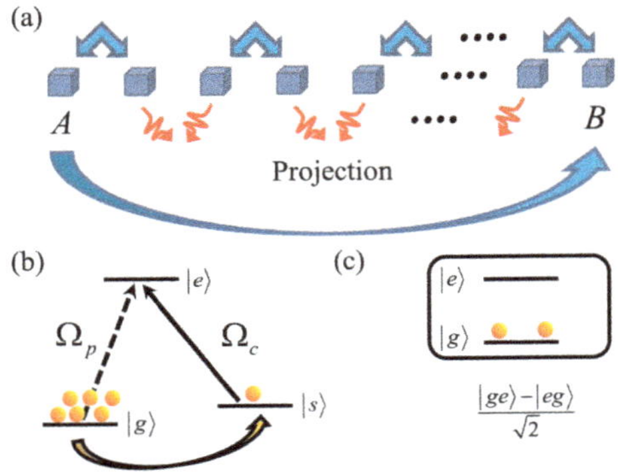

Figure 7.1. A quantum repeater protocol, EIT setup, and two-level subradiant states. (a) The quantum repeater distributes quantum entanglement between unit A and B by inserting entangled pairs (cubes) denoted by double arrows. Projection is done by Bell measurements denoted by wavy arrows. (b) Three atomic levels in the EIT setup where a probe field Ω_p is absorbed and transferred to a collective spin excitation in $|s\rangle$ with a presence of control field Ω_c. (c) Two-level quantum emitters forming an antisymmetric subradiant state.

forms a basic unit of quantum memory that can switch the quantum information between light and atoms. Another useful scheme is simply two-level atoms, where phase-imprinted subradiant states can be formed and manifest a long lifetime [11, 12]. We note that there are other techniques in constructing quantum memory, for example of using gradient echo memory and atomic or optical frequency comb [7], which will not be discussed here.

For a last remark in this section, we note of an exponential improvement in EIT memory [13] when the atoms are coupled with a nanofiber under a tailored control field. This shows the use of subradiant states to suppress the free-space decay and at the same time to increase the coupling rate to the guided modes in the fiber. Another demonstration of using subradiant modes is to reflect light completely in two-dimensional quantum emitters [14]. These two examples present the capability of collective interactions within dense quantum emitters, which can be harnessed to change the property of light propagation and even to go beyond the limitation of storage errors or the bound of storage efficiency in the conventional single-particle perspective of atom–light interactions.

7.1.2 Quantum state engineering

Here I discuss the quantum state engineering as one of quantum information applications. This is essential in demonstrating the control and manipulation of quantum states. A somewhat overlapped concept of state preparation particularly refers to an initial or final state of interests. The state engineering actually focuses on the techniques to switch between various quantum states, which accesses different properties of these states, for example symmetries, collective decay rates, or even topologies.

In figures 3.2 and 3.3, I have demonstrated a phase imprinting technique to manipulate super- and sub-radiant states in either singly- or multiply-excited Hilbert spaces. This can be done by applying external magnetic field gradients or light pulses to imprint AC-Stark shifts. This however is not perfect and is limited by the lifetime of quantum states. Therefore, the operation of imprinting required phases will be deteriorated in its efficiency. A way to go around this lifetime limitation is to use the EIT setup introduced in the previous section, where a spin excitation can serve a qubit with sufficient lifetime. Another candidate is Rydberg transition, which naturally has a longer intrinsic decay time for higher principal quantum numbers. This would empower the EIT setup with a Rydberg-EIT [15], where the state $|s\rangle$ is replaced by the Rydberg atomic level. A bonus comes from this Rydberg transition, which further offers strong dipole–dipole interactions via a blockade effect from high-lying Rydberg levels and leads to phase gate operations essential to quantum computation.

Recent progress involves individual manipulations of atom assembler using optical tweezers [16, 17], which is expected to further enable individual state engineering of neutral atoms. This promises large-scale quantum engineering and paves the way toward a practical quantum processor to outperform classical computers. Many other platforms are competing to achieve this feat, including ions, defect centers in diamonds, or superconducting qubits, and yet none of them claim they are unparalleled. We are fortunate to live in this racing era, where every platform of quantum systems has a chance to stand out.

I take one more example of superconducting qubits, which recently show a controlled switching between super- and sub-radiant states [18]. This demonstrates a coherent control on the collectively excited state in a circuit system. However, an expected complexity in these qubits will arise quickly when their number surpasses ten or even a hundred of them. An issue of scalability always hinders the potentiality of quantum systems, since they are fragile to perturbations even a tiny one. One manifestation of this fragility can be shown in the performance of a 'quantum radar' [19] in quantum metrology, where one photon loss in N maximally entangled photons results in a catastrophe of quantum enhancement. Nonetheless, to reach quantum supremacy actually costs a lot, but it is worth continuing to pursue this for its unprecedented capability.

7.2 Quantum simulation

A quantum simulator originally proposed by Feynman is to resolve the complexity of simulating quantum systems by classical computers. If we can simulate a quantum phenomena via another quantum system, where we control its Hamiltonian well, a quantum simulator can then take a partial role of a quantum computer yet to exist, to tackle a task that classical computers might take an exponentially long time on system sizes. In principle, one can fine-tune the Hamiltonian close to the system of interests and predict the physical observables under appropriate measurements [20]. With controllable light–matter interactions to tailor system Hamiltonians, quantum simulations can be implemented, for example, in ultracold quantum gases [21], trapped ions [22], or photonic systems [23].

To summarize the basic requirements in a quantum simulator to have its supremacy as in quantum computers, we need first a quantum system of particles, fermions or bosons, and a series of control techniques in state initializations,

Hamiltonian engineering, detection, and verification [20]. Here I will not address the details of these techniques that are essential and broad enough to deserve another book for discussion. However, I focus on how new systems of particles arise and become candidates for quantum simulators via controlling the collective interactions between light and atoms. Below I discuss many-body spin dynamics by tailoring long-range dipole–dipole interactions and simulations of interesting steady-state phases driven by this collective interaction.

7.2.1 Many-body spin dynamics

In this section, I first focus on a platform of the photonic waveguide crystal, where the guided modes of light couple with distant trapped atoms to form long-range atom–atom interactions [24]. This further leads to a tunable spin-exchange interaction that can realize an effective *XY* and two-dimensional *XXZ* Hamiltonians, with the help of auxiliary pump fields. In addition to this versatile control over Hamiltonians in a nanophotonics platform, it can potentially simulate nonequilibrium quantum dynamics with strong and long-range spin interactions and unveil the role of interaction in thermalization of systems.

As another example, a quantum nonlinear system effectively provides a many-body state of light instead. With effective photon–photon interactions mediated by atomic Rydberg states, three-photon and two-photon bound states are recently observed [25]. This shows in bunching photon–photon correlations, which leads to the effective photonic trimers and dimers. This nice experiment demonstrates a new and potential platform to simulate exotic quantum states in photons.

The above two demonstrations present the power of controlled interactions, which makes quantum simulations of strongly correlated systems and exotic many-body phases of matter or light possible, respectively. As a last remark here, I specifically discuss the potentiality of quantum gases in quantum simulations, which are best in high degrees of controllability [21]. With the capability of tuning atom–atom interactions, the ultracold Fermi gases can simulate a BEC (Bose–Einstein condensate) to BCS (Bardeen–Cooper–Schrieffer) superfluid crossover, when the interactions are tuned from a weakly repulsive to attractive side. With controlled on-site interactions in optical lattices, a weakly interacting BEC simulates a Bose–Hubbard model and presents a superfluid to Mott insulator transition, where matter wave correlations are long-ranged and exponentially suppressed, respectively. Finally, artificial gauge fields can be generated to simulate the quantum Hall effect with the aid of laser-assisted hopping between neighboring sites in optical lattices. These controlled techniques of Hamiltonians and interactions within can further exploit long-range interactions of Rydberg atoms or polar molecules to investigate nonergodic many-body dynamics and realize an alternative quantum computer [21].

7.2.2 Interaction-driven steady-state phases and nonergodic dynamics

In this last subsection, I introduce one quantum optical system of chirally coupled quantum emitters [26], which can simulate many interesting steady-state phases [27]. The main results of the driven-dissipative phases have been shown in section 6.4, where the distributions of excited-state populations present crystalline orders and edge or hole excitations. In this particular quantum optical system with non-reciprocal couplings, the steady-state phases can be interaction-driven, which can be tuned by waveguide mediated dipole–dipole interactions determined by interatomic distances.

The simulated phases of quantum matter actually depend on the controllability of interactions. I have mentioned the BEC–BCS crossover in the previous subsection, where a short-range scattering length can be tuned in Fermi gases. In Rydberg states or polar molecules, long-range interactions depend on the principal quantum numbers or excitation configurations respectively, where a supersolid phase can be formed. In chirally coupled quantum emitters, the dipole–dipole interactions are mediated by the guided modes which allow an infinite-range coupling. Many exotic many-body states can then be realized and simulated in many platforms of quantum interface between light and atoms, which enables new phases of matter that conventional condensed matter systems cannot easily access.

For a last remark, in figures 7.2 and 7.3, I show two examples of nonergodic system dynamics that lasts for a very long time in a system of chirally coupled

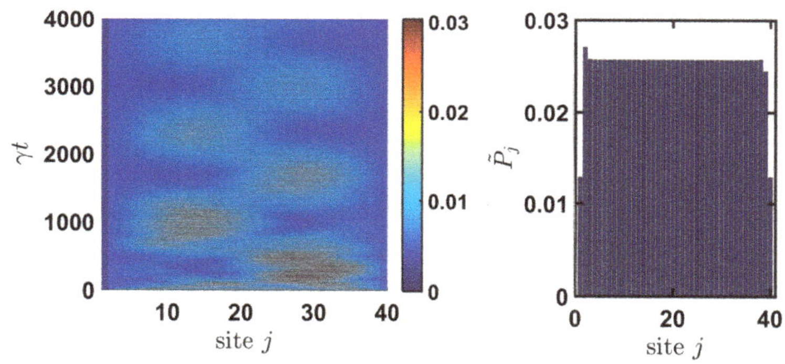

Figure 7.2. Nonergodic dynamics and steady state with hole excitations. The parameters are chosen at $D = 0.1$ and $\xi = 2\pi/4$ with $N = 40$.

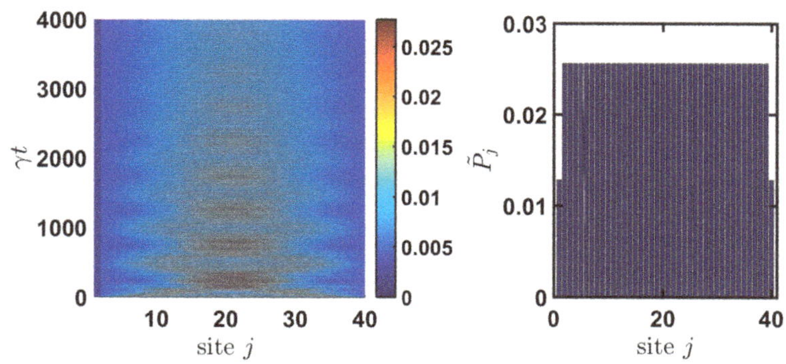

Figure 7.3. Nonergodic dynamics and steady state with hole excitations. The parameters are chosen at $D = 0.0$ and $\xi = 2\pi/4$ with $N = 40$.

quantum emitters. The right panels in the respective figures show the normalized distributions of steady-state populations, which present the phase of bi-hole excitations at the boundary of the atomic chain. The respective left panels demonstrate how the system relaxes to the steady states, which takes very long time (at least more than 10^3 of γt), longer than its intrinsic lifetime $1/\gamma$. When D is finite (see its definition in section 6.2) in figure 7.2, the atomic excited-state populations start to transport towards the right initially and bounces back and forth until they reach the pattern in the right panel. This shows a similar pattern of Newton's cradle that is only allowed in integrable quantum systems [28]. In contrast, in figure 7.3, a different pattern shows up under reciprocal couplings, where left- and right-propagating decay rates are the same. This pattern is symmetric to the center of the atomic chain, where most of the excited-state populations occupy around the center. These long-time dynamics present the nonthermalizing features, and therefore a chirally coupled atomic chain with well-controlled interaction parameters can be suitable for studying nonequilibrium physics.

References

[1] Nielsen M A and Chuang I L 2000 *Quantum Computation and Quantum Information* (Cambridge: Cambridge University Press)

[2] Bouwmeester D, Ekert A K and Zeilinger A 2000 *The Physics of Quantum Information: Quantum Cryptography, Quantum Teleportation, Quantum Computation* (Berlin: Springer)

[3] Hammerer K, Sørensen A S and Polzik E S 2010 Quantum interface between light and atomic ensembles *Rev. Mod. Phys.* **82** 1041–93

[4] Lin Y, Leibrandt D R, Leibfried D and Chou C W 2020 Quantum entanglement between an atom and a molecule *Nature* **581** 273–7

[5] Acín A *et al* 2018 The quantum technologies roadmap: a European community view *New J. Phys.* **20** 080201

[6] Brooks M 2019 Beyond quantum supremacy: the hunt for useful quantum computers *Nature* **574** 19–21

[7] Cho Y W *et al* 2016 Highly efficient optical quantum memory with long coherence time in cold atoms *Optica* **3** 100–7

[8] Heinze G, Hubrich C and Halfmann T Stopped light and image storage by electromagnetically induced transparency up to the regime of one minute *Phys. Rev. Lett.* **111** 033601

[9] Dudin Y O, Li L and Kuzmich A 2013 Light storage on the time scale of a minute *Phys. Rev.* A **87** 031801

[10] Fleischhauer M, Imamoglu A and Marangos J P 2005 Electromagnetically induced transparency: optics in coherent media *Rev. Mod. Phys.* **77** 633–73

[11] Jen H H, Chang M S and Chen Y C 2016 Cooperative single-photon subradiant states *Phys. Rev.* A **94** 013803

[12] Jen H H 2017 Phase-imprinted multiphoton subradiant states *Phys. Rev.* A **96** 023814

[13] Asenjo-Garcia A, Moreno-Cardoner M, Albrecht A, Kimble H J and Chang D E 2017 Exponential improvement in photon storage fidelities using subradiance and 'selective radiance' in atomic arrays *Phys. Rev.* X **7** 031024

[14] Shahmoon E, Wild D S, Lukin M D and Yelin S F 2017 Cooperative resonances in light scattering from two-dimensional atomic arrays *Phys. Rev. Lett.* **118** 113601

[15] Saffman M, Walker T G and Mølmer K 2010 Quantum information with Rydberg atoms *Rev. Mod. Phys.* **82** 2313–63
[16] Barredo D, de Léséleuc S, Lienhard V, Lahaye T and Browaeys A 2016 An atom-by-atom assembler of defect-free arbitrary two-dimensional atomic arrays *Science* **354** 1021–3
[17] Endres M *et al* 2016 Atom-by-atom assembly of defect-free one-dimensional cold atom arrays *Science* **354** 1024–7
[18] Wang Z *et al* 2020 Controllable switching between superradiant and subradiant states in a 10-qubit superconducting circuit *Phys. Rev. Lett.* **124** 013601
[19] Maccone L and Ren C 2020 Quantum radar *Phys. Rev. Lett.* **124** 200503
[20] Cirac J I and Zoller P 2012 Goals and opportunities in quantum simulation *Nat. Phys.* **8** 264
[21] Bloch I, Dalibard J and Nascimbène S 2012 Quantum simulations with ultracold quantum gases *Nat. Phys.* **8** 267
[22] Blatt R and Roos C F 2012 Quantum simulations with trapped ions *Nat. Phys.* **8** 277
[23] Aspuru-Guzik A and Walther P 2012 Aspuru-Guzik A and Walther P 2012 Photonic quantum simulators *Nat. Phys.* **8** 285
[24] Hung C L, González-Tudela A, Cirac J I and Kimble H J 2016 Quantum spin dynamics with pairwise-tunable, long-range interactions *Proc. Natl. Acad. Sci.* **113** E4946–55
[25] Liang Q Y *et al* 2018 Observation of three-photon bound states in a quantum nonlinear medium *Science* **359** 783–6
[26] Jen H H, Chang M S, Lin G D and Chen Y C 2020 Subradiance dynamics in a singly excited chirally coupled atomic chain *Phys. Rev.* A **101** 023830
[27] Jen H H 2020 Steady-state phase diagram of a weakly driven chiral-coupled atomic chain *Phys. Rev. Res.* **2** 013097
[28] Kinoshita T, Wenger T and Weiss D S 2006 A quantum Newtonas cradle *Nature* **440** 900

IOP Publishing

Collective Light Emission

Many quantum emitters

Hsiang-Hua Jen

Chapter 8

Future perspectives

We discuss several future research directions and their challenges in the subject of collective light emissions from many quantum emitters. We focus on the rising complexity when more emitters are involved, the requirement of a strong coupling regime for interesting phenomena to appear, the exotic collective states, and many-body phenomena where distinct behaviors of light or atoms can emerge.

8.1 Complexity of a macroscopic system

Emergent phenomena are highly related to the complexity of a system, either for a classical or quantum mechanical one. The complexity results from a system's multiple scales in space and time, which makes the understanding of a system difficult and even intractable when the entities or constituents of a system become macroscopic. What is macroscopic is actually hard to define and is often not clearly differentiated from what is microscopic. This difficulty is also reflected in the emergent phenomena when a system is growing in size, which can be attributed to 'more is different' [1] due to the hierarchical structure of the physical world.

To most we can only distinguish a noninteracting system from an interacting one, and it is the interaction within a system that matters. Complexity arising from interactions and their nonlinearity results in system's fluctuations, multiple scales, and emergent structure [2]. In figure 8.1, multiple scales of space and time is shown schematically in collective emissions. The scale of space can be approximately categorized by as small as the interatomic distances r and as large as the boundary of a whole ensemble of atoms L. I use dots between r and L to show that other emergent space scales can exist if mutual correlations are established in an intermediate region. The interatomic distances determine how strong the atom–atom interactions are, while the macroscopic scale or simply the geometry of the atoms decides the properties of light emissions as well. We note that even under the same density, the arrangement of these atoms also plays a role, for example they can be periodically or randomly distributed. Additionally, a gray parameter regime

doi:10.1088/978-0-7503-2699-5ch8

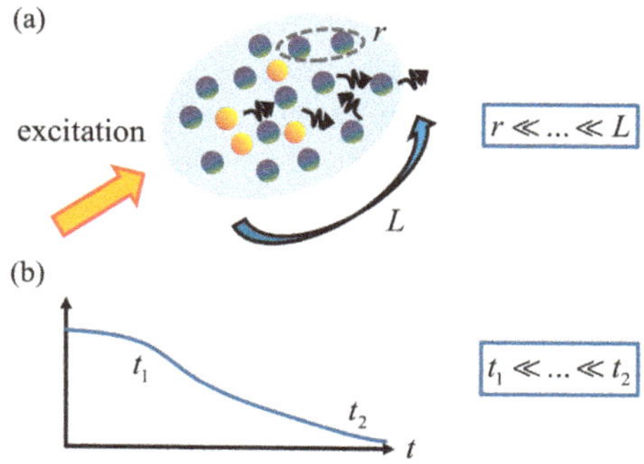

Figure 8.1. Schematic demonstration of multiple scales of space and time in collective light emission. (a) When an ensemble of atoms are excited, multiple scattering events denoted by thick arrows can happen. In the scattering process, a macroscopic scale of ensemble length L is contrasted by interatomic distance r. (b) After the excitation, a spontaneous emission may involve multiple timescales.

exists of how large N (number of atoms) is large enough to consider a continuous limit of system interactions and the discrete nature of constituents can be safely removed. We note a nice discussion in comparing the Beer–Lambert law conventionally accepted in a bulk system on the absorption of light, which can be violated in an optically thin two-dimensional gas [3] when multiple scattering events are crucial under strong dipole–dipole interactions.

Similarly, the collective emissions in figure 8.1(b) present several timescales in an initial, middle stage, and the tails of the decay, in contrast to a simple exponential decay of spontaneous emission from the independent atoms. The complexity of these multiple timescales can be attributed to many excited eigenmodes supported by the system, which is more significant in the subradiant emissions where different subradiant modes with different characteristic lifetimes emerge. This multiple timescale, however, is not obvious in the superradiant regime since all excitations decay quickly before sufficient populations sustain to show a different decay behavior. In a nice demonstration of an atom cloud [4], a different timescale shows up after the system is superradiantly relaxed, and this subradiant timescale prolongs as the density of the atom cloud increases. It is extraordinary to present this different timescale, which provides evidence of the existence of subradiant modes and shows the capability to detect weak signals of order $\sim 10^{-3}$ and lower of its initial maximum.

To close the discussion on the complexity of a macroscopic system, I use the conclusion on the observation of coherent scattering from a dense atomic cloud [5]:

> 'As a conclusion, we have measured the coherent scattering by a dense, cold atomic cloud. We have observed a failure of standard models, such as the FHM model or the mean-field Lorentz model. The remaining difference with the microscopic model shows that a quantitative understanding of the light-induced interactions even in a relatively simple situation is still a challenge.'

The FHM (Friedberg, Hartmann, and Manassah) model [6] refers to a mean-field approach which approximates the local field by including the averaged fields from the medium other than the local one. This leads to a Lorentz–Lorenz or Clausius–Mossotti formula [7] as an effective correction to the noninteracting case. It is called standard since the mean-field approach has successfully calculated light properties in a slab geometry in an early time, inspired by electromagnetism. However, when quantum emitters are cold enough that the Doppler effect does not play a role as in conventional homogeneous media, the detailed atom–atom correlations induced by dipole–dipole interactions become dominant and a mean-field treatment is insufficient to account for these correlations. On the other hand, the light-induced atom–atom correlations can be included and is denoted as a microscopic model quoted above, where its interaction form in free space has been introduced in chapter 2. Unfortunately, the comparisons between the microscopic model and the experimental observations still show mismatches in both linewidth and central frequency of the scattering spectrum.

It looks like the theoretical models face quite a challenge 'even in a relatively simple situation' [5]. A challenge could also mean an opportunity to uncover a new mechanism that could have been missed in fundamental light–matter interactions. Here I raise some points of considerations that could be useful for further investigation of light scattering from this simple setup of an atom cloud.

First, the so-called microscopic model actually derives from a more fundamental master equation of light interacting with atoms [8] and is obtained in an operational or Lindbladian form by tracing out the field degrees of freedom. We can take one step back without the field tracing and instead consider a field propagation of a quantized light through the medium. This way we can reveal the role of quantum fluctuations, which could be crucial especially when a quantized light is scattered. The propagation of a quantized light can be put into quantum Langevin equations with quantum noises, but they are valid only when the number of atoms $N \gg 1$. For a small sample with a moderate N, the propagation effect can be neglected, or quantum fluctuations can still be important to some degree before the quantum Langevin approach breaks down. This would be the most difficult parameter regime for theoretical investigations to directly compare with experiments.

Secondly, a microscopic model becomes self-consistent only when a truncation is made to their hierarchical coupled equations. Conventionally, a truncation to the first or second order of correlations in the model is enough, especially under a perturbation approximation for a weak field excitation. Since the microscopic model considered [5] is truncated to the first order of atomic operators, a natural conjecture of improving the theoretical model is to include the next order correlation into the model. It could be again negligible but is worth trying to prove so. Furthermore, an inclusion of second-order correlations is nontrivial, owing to the enlarged space of coupled equations. For example, the correlations of two ground or one ground and one excited state, and two coherence operators $\sigma_\mu^\dagger \sigma_\nu$ would emerge. This consideration of higher correlations becomes even more daunting when multi-photon excitation is considered. This leads to the third possible improvement in theory by including a multi-excitation Hilbert space. A multi-photon scattering within many atoms again is nontrivial, and much larger (exponentially increasing) Hilbert space would be required and easily suffocate the computation power. If higher correlations are involved, we will further need to take care of correlations in, for example, two excited states or two coherence operators $\sigma_\mu^\dagger\sigma_\nu^\dagger$.

Lastly, a better and precise control of experimental parameters would be beneficial in a close comparison between theory and experiment. This involves a better characterization of sample sizes, atomic density, and excitation light property, and a clarification of the effect of multiple atomic levels. An optical pumping can be used to prepare the system into a genuine two-level configuration, and we expect that a better design of measuring quantized light scattering in a dense and inhomogeneous media can provide more insight into the roles of quantum fluctuations.

8.2 Strong coupling regime

A strong coupling regime shows in many aspects of an ideal quantum interface of light and atoms. It indicates a higher efficiency of quantum interface or a genuine realization of many-body states in light-mediated quantum systems. In the context of collective emissions from many quantum emitters, a strong coupling regime means strong resonant dipole–dipole interactions that significantly present superradiant or subradiant radiation. This strong coupling regime is reached when the interatomic distance is much less than the excitation wavelength. Generally speaking, a strong coupling regime can be achieved in many platforms, and the essence of it relies on a strong confinement of light field which efficiently couples to the medium [9]. For example, in figure 8.2, it can done by placing the atoms inside a cavity, using an atom cloud with a large optical depth, putting a quantum dot on an optical nanofiber [10] or in a photonic crystal waveguide [11]. The essence of strong coupling regime relies on a strong confinement of light field which efficiently couples to the medium.

As an example, I use an atom–waveguide system with a one-dimensional light-induced dipole–dipole interaction. We can use a factor of

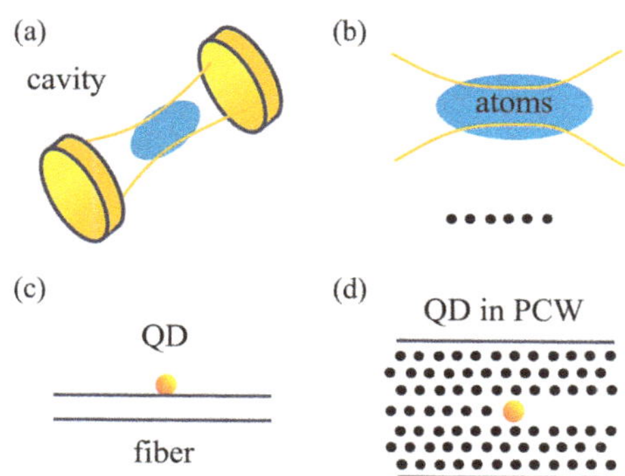

Figure 8.2. Schematic plots for various strong coupling regimes: (a) cavity QED, (b) free-space atoms with a large optical depth or interacting with a tightly confined light, (c) quantum dot (QD) on a nanofiber and (d) quantum dots in a photonic crystal waveguide (PCW).

$$\beta \equiv \frac{\gamma_L + \gamma_R}{\gamma_L + \gamma_R + \gamma_{non}}, \tag{8.1}$$

to characterize how strong the coupling is. γ_{non} denotes the nonradiative losses or nonguided mode couplings. If we assume $\gamma_L + \gamma_R = \gamma$ as an effective total coupling rate in the guided modes, a $\gamma_{non}/\gamma \sim 0.1$ leads to $\beta \sim 90\%$, while a $\gamma_{non}/\gamma \sim 0.01$ leads to $\beta \sim 99\%$. This γ_{non} characterizes a time window of $1/\gamma_{non}$ for any phenomena to be observable. It is particularly crucial for subradiant dynamics [12] that lasts for long time and is very fragile to nonradiative modes.

Another equivalent factor of cooperativity

$$\eta \equiv \frac{\gamma_L + \gamma_R}{\gamma_{non}}, \tag{8.2}$$

can also give a measure for strong coupling regime when $\eta \gg 1$. As a consequence, $\eta \sim 10$ and 100 when $\gamma_{non}/\gamma \sim 0.1$ and 0.01 respectively.

We note that β can exceed 90% [13] or 98% [11] in a system of quantum dots in photonic crystal waveguides. However, it could be challenging to create an array of identical quantum dots, but this can be overcome by tuning inter-dot tunneling rates to control over individual quantum dots. Other promising solid-state systems involve superconducting qubits which can also be manipulated in an individual qubit level. For neutral atoms trapped around an optical nanofiber, we note recent progress in creating a singly collective excitation [14], which offers the basic element in quantum network and paves the way towards a large-scale and strongly-coupled waveguide QED system.

Much ongoing effort has been put into enhancing the light–matter couplings to the strong coupling regime. This is essential not only in generating an efficient quantum interface, creating a genuine entangled state, building up quantum nodes to store and control the quantum information, but also crucial in investigating quantum many-body phenomena. In the promising future, a quantum computer would come to this realization by taking these advances in various continuously improved platforms under strong couplings.

8.3 Exotic collective states and many-body phenomena

In Chapter 4, I mentioned the fermionic features of doubly-excited subradiant states [15, 16], and relevant twilight states [17] that show strong and long-lived photon–photon correlations. These studies open up new directions other than superradiant or subradiant emissions from many quantum emitters and reveal the many-body features of doubly-excited subradiant states. Actually there is a broad intermediate region to show collective emissions with moderate super- or sub-radiant rates. The twilight states, for example, only involve a product state of singly-excited bright and subradiant states. This is already interesting enough to be explored and shows the richness of light–matter interacting systems, which can be harnessed to tailor the radiation properties of constituent photons.

Further investigation can extend to many other collective states more than two excitations [17]. This is indeed a large pool of collective states under a multiply-excited Hilbert space. It seems that to look for some useful collective state is like finding a needle in a haystack. However, we can start from the bottom up. It is already nontrivial to look for the triply-excited case, where either doubly-excited bright with singly-excited subradiant or singly-excited bright with doubly-excited subradiant states could be identified. This will lead to an investigation of a bound photonic state of two excitations, which would be useful for an application of storage of multiple photons and also fundamental in studying many-body states of light.

We note a recent demonstration of creating Laughlin states of light [18], which shows the capability of simulating fractional quantum Hall effect through interacting dark-state polaritons in the Rydberg-EIT setup and a twisted cavity with Floquet engineering [19]. It is the capability of using the light–matter interface to create nonlinear interactions, along with new techniques to manipulate the properties of light, which further enables an engineering of the correlated states to present topological properties. These collective states present a new kind of synthetic quantum matter, which can be controlled and manipulated on the level of interaction energy or even Hamiltonians, leading to a very bright future of quantum simulations using only neutral atoms and light.

References

[1] Anderson P W 1972 More is different *Science* **177** 393–6

[2] Katherine W 2019 Life is physics *Am. Phys. Soc. Phys.* **12** 2

[3] Chomaz L, Corman L, Yefsah T, Desbuquois R and Dalibard J 2012 Absorption imaging of a quasi-two-dimensional gas: a multiple scattering analysis *New J. Phys.* **14** 055001

[4] Guerin W, Araújo M O and Kaiser R 2016 Subradiance in a large cloud of cold atoms *Phys. Rev. Lett.* **116** 083601

[5] Jennewein S *et al* 2016 Coherent scattering of near-resonant light by a dense microscopic cold atomic cloud *Phys. Rev. Lett.* **116** 233601

[6] Friedberg R, Hartmann S R and Manassah J T 1973 Frequency shifts in emission and absorption by resonant systems of two-level atoms *Phys. Rep.* **7** 101–79

[7] Zangwill A 2013 *Modern Electrodynamics* (Cambridge: Cambridge University Press)

[8] Lehmberg R H 1970 Radiation from an N-atom system. I. General formalism *Phys. Rev.* A **2** 883–8

[9] Lodahl P, Mahmoodian S and Stobbe S 2015 Interfacing single photons and single quantum dots with photonic nanostructures *Rev. Mod. Phys.* **87** 347–400

[10] Yalla R, Sadgrove M, Nayak K P and Hakuta K 2014 Cavity quantum electrodynamics on a nanofiber using a composite photonic crystal cavity *Phys. Rev. Lett.* **113** 143601

[11] Arcari M *et al* 2014 Near-unity coupling efficiency of a quantum emitter to a photonic crystal waveguide *Phys. Rev. Lett.* **113** 093603

[12] Jen H H, Chang M S, Lin G D and Chen Y C 2020 Subradiance dynamics in a singly excited chirally coupled atomic chain *Phys. Rev.* A **101** 023830

[13] Söllner I *et al* 2015 Deterministic photon-emitter coupling in chiral photonic circuits *Nat. Nanotechnol* **10** 775–8

[14] Corzo N V, Raskop J, Chandra A, Sheremet A S, Gouraud B and Laurat J 2019 Waveguide-coupled single collective excitation of atomic arrays *Nature* **566** 359–62

[15] Albrecht A, Henriet L, Asenjo-Garcia A, Dieterle P B, Painter O and Chang D E 2019 Subradiant states of quantum bits coupled to a one-dimensional waveguide *New J. Phys.* **21** 025003

[16] Zhang Y X and Mølmer K 2019 Theory of subradiant states of a one-dimensional two-level atom chain *Phys. Rev. Lett.* **122** 203605

[17] Ke Y, Poshakinskiy A V, Lee C, Kivshar Y S and Poddubny A N 2019 Inelastic scattering of photon pairs in qubit arrays with subradiant states *Phys. Rev. Lett.* **123** 253601

[18] Clark L W, Schine N, Baum C, Jia N and Simon J 2020 Observation of Laughlin states made of light *Nature* **582** 41–5

[19] Clark L W, Jia N, Schine N, Baum C, Georgakopoulos A and Simon J 2019 Interacting Floquet polaritons *Nature* **571** 532–6

Lightning Source UK Ltd.
Milton Keynes UK
UKHW051920301020
372498UK00003B/61